Troubled Fields

Eric Ramírez-Ferrero

Columbia University Press   New York

Troubled Fields

MEN, EMOTIONS, AND THE CRISIS
IN AMERICAN FARMING

COLUMBIA UNIVERSITY PRESS
Publishers Since 1893
New York   Chichester, West Sussex
Copyright © 2005 Columbia University Press
All rights reserved
All photos courtesy of The Museum of The Cherokee Strip,
   Oklahoma Historical Society
Library of Congress Cataloging-in-Publication Data
Ramírez-Ferrero, Eric, 1963–
   Troubled fields : men, emotions, and the crisis in American farming / Eric
Ramírez-Ferrero.
      p. cm.
   Includes bibliographical references and index.
   ISBN 0–231–13024–4 (cloth : alk. paper) — ISBN 0–231–50363–6 (electronic) —
ISBN 0–231–13025–2 (paper : alk. paper)
   1. Suicide—Oklahoma. 2. Farmers—Oklahoma—Psychology. 3. Agriculture—
Economic aspects—Oklahoma. I. Title.
HV6548.U52057 2005
362.28′1′0886309766—dc22

                                                                    2004051981

Columbia University Press books are printed on permanent
   and durable acid-free paper.
Printed in the United States of America
Designed by Lisa Hamm
c 10 9 8 7 6 5 4 3 2 1
p 10 9 8 7 6 5 4 3 2 1

A la memoria de mis padres,

M. Luisa Ramírez Ferrero y Justo José Ramón Ramírez

Y a mi hermano, Nelson

Mientras más me busco, más los encuentro . . .

CONTENTS

ACKNOWLEDGMENTS

I BELIEVE researchers often choose their subjects of study not only because they address central intellectual concerns, but also because the topics resonate at a personal level. For me, research in my home state provided the opportunity to explore my own sense of "community." What I found while conducting fieldwork was that my community was there all along in the form of supportive mentors, colleagues, and friends.

First, I am deeply grateful to my dissertation committee members, both formal and informal. Carol Delaney served as my principal advisor and continues to serve as an exemplar of the type of feminist scholarship to which I aspire. Bernard Siegel recently passed away; I will greatly miss his intellectual guidance and avuncular personal support, which made this project possible. Howard Stein's work was a major source of inspiration for this study, as is his dedication to the health of rural Oklahomans. Finally, Peter Stromberg's incisive critiques of the manuscript were fundamental to the development of the arguments presented here.

I am also grateful to Clifford Barnett and Nancy Scheper-Hughes for serving on my doctoral committee and for their commitment to a problem-focused, applied anthropology that makes a difference. They have inspired me more than they will ever know. I appreciate the support and inspiration of Jane Fishburne Collier and Sylvia Junko Yanagisako, whose influence on my thinking I hope will be apparent throughout this book. I am grateful to Ruth Behar for encouraging me to write about my fieldwork experiences and explore my own sense of "community."

A special thank-you is warranted to my first anthropology teachers: Ethel Vesper, for making me fall in love with anthropology, and Joanna Kirkpatrick, for disciplining my new passion, giving it form and voice.

I owe a debt of gratitude to the staff of Rural Health Projects, especially Jackie Longacre, Richard Perry, and Mary Jac Rauh, for their material and personal support, which enabled me to conduct research and synthesize and analyze my data. I also want to acknowledge my "partners in crime," Lucia Rojas-Smith and Sydney Kriter, who made tooling around all over northwestern Oklahoma conducting focus group research and "first responder" trainings a joy.

Many thanks to the senior management team at Planned Parenthood of Arkansas and Eastern Oklahoma for granting me time away from work to write. Staff members include Nancy Kachel (President and CEO), Sue Riggs, Xan Blake, Becky Collins, Laurie Smith, and Pat Kroblin. I also want to thank my former staff in Tulsa and Little Rock, especially Sheila Lugene Asher, for keeping the education department running flawlessly in my absence.

Many friends provided instrumental support toward the completion of this project. I would like to offer my appreciation to Pat Bellmon, who is from a long line of Oklahoma farmers. Her encouragement, friendship, and good counsel were vital in helping me develop key arguments presented in this study. I also want to thank Claire Moor Aaronson for her beautiful translation of the Torres Bodet poem that appears in the appendix; Diana Aaronson for her careful readings of the manuscript and her thoughtful comments; Richard L. Phillips for a beautiful setting in which to think and write the chapter "The Good Farmer"; Debi Sanditen and Laura Belmonte, my writing buddies, who prodded me on to finish the manuscript by encouraging me to join them at cafés or libraries when they had writing projects of their own to complete. Though perhaps I did not show my appreciation at the time (ahem . . . ahem . . .), I am sure the incessant goading of my dear friends Robert Inglish and Craig Wood played a major role in this book's completion.

I appreciate the efforts of Barry Bloyd, state statistician for the Oklahoma Department of Agriculture, who made available and provided preliminary analysis of data about the changing structure of agriculture in Oklahoma, a theme that is central to this book.

Glen McIntyre of the Museum of the Cherokee Strip in Enid, Oklahoma, opened his doors to me, allowing me to scour the Museum's archive of photographs. His staff, especially Jennifer Jones and Jack Taylor, skillfully winnowed the options, saving me an enormous amount of time and effort. They provided advice on picture choices and made sure the selected art would be of publication quality.

I want to thank Jane Adams and an anonymous reviewer at Columbia University Press for their careful reading of the manuscript and insightful responses. I believe the integration of their suggestions has made this a stronger study.

I want to express my gratitude to Wendy Lochner, my editor, and the rest of the staff of Columbia University Press, especially Robert Fellman, Michael Haskell, Anne McCoy, and Suzanne Ryan. Wendy has guided this project with aplomb. In the process she has been supportive, encouraging, and a joy to work with.

I am very fortunate to have a wonderful and amazingly supportive family who make possible everything I do. They include Eric Flegel; Nelson, Johna, and Liana Ramírez (my smart and beautiful niece) and the rest of the Ramírez and Ferrero families in Florida, Cuba, and beyond; Stanley, Diana, and Claire Aaronson; Deborah, Justin, Jordan (my precocious godson), and Alex Cromwell; Cinde Donoghue and Pete Dowty, Michael Evanish and Jeff Windus.

Finally, I reserve my most heartfelt gratitude to the men and women who participated in this study. Their generosity was humbling and their hospitality, a gift. I feel honored to have been allowed to listen to stories about their lives and their experiences resulting from the ongoing modernization of the agricultural sector. I hope they feel that, with this work, I have done justice to their stories. I also hope that this book can make even some small difference toward the betterment of rural life in my home state and elsewhere.

I remember that when I could not make sense of the cultural material I was gathering, I would run to anthropological theory for cover and help. One mentor, Howard Stein, suggested I not do that. Rather, he said, "Listen carefully to what farmers are telling you. It's all there; you just have to listen more closely." Out of a deep sense of respect and thanksgiving to the men and women with whom I worked, I hope I listened well enough.

Troubled Fields

There is a sense in which rapid economic progress is impossible without painful adjustments. Ancient philosophies have to be scrapped; old social institutions have to disintegrate; bonds of caste, creed and race have to burst; and large numbers of persons who cannot keep up with progress have to have their expectations of a comfortable life frustrated. Very few communities are willing to pay the full price of economic progress.

—United Nations, Department of Social and Economic Affairs, *Measures for the Economic Development of the Underdeveloped Countries*, 1951

INTRODUCTION

Homework

MY BOSS and I had visited the state Department of Agriculture in Oklahoma City to survey data about the health of farming communities in northwestern Oklahoma. It was 1991 and our agency had recently contracted with the Oklahoma State University Cooperative Extension Service to assess and address the mental health needs of rural residents in light of the ongoing financial crisis in family farming.

We met with Rick,[1] a man in his early thirties, who was a data specialist in the department. Our conversation was jovial and he was clearly eager to support our efforts. He shared several useful reports his division had produced, including transcripts from hearings held around the state documenting the social impact of the farm crisis.

As our attention shifted from quantitative data to stories, however, the texture of our conversation changed. It became more weighted and somber as he told us of his professional work on the issue. Rick described the letters his department received from desperate farmers seeking answers—or redress for abusive practices perpetrated, they allege, by unscrupulous lenders. He recounted his moving experiences working with the Governor's Task Force on the Rural Farm Crisis, learning firsthand about its social and economic complexity and its painful consequences, transforming rural communities—and people—in fundamental ways.

But he also spoke to us about his personal experiences: his family's long-term struggle with farm financial indebtedness and the health consequences of this process, including his father's recent psychiatric hospitalization for attempted suicide.

"It's not over," Rick said. "People think that the farm crisis is something that happened in the early eighties, that the suicides and all are over, but it's not. It's not."

Then he added, "You know, someone really should look into the crisis . . . the suicides . . . what's happening now."

Jackie, my boss, flashed me her by now familiar look, an expression that meant, "Take note. This is important." In this particular case, I knew it also implied that the "someone" who should take on this project was me.

I didn't need convincing. My public health work in rural communities, together with frequent media reports regarding the farm crisis, had already galvanized my interest and concern. One story, in particular, had riveted my—and the state's—attention. Collaborating on a report about farm accidents, traditionally thought to be the leading cause of agricultural fatalities, investigators from the *Tulsa World* and Oklahoma State University (OSU) made a surprising discovery during the course of their investigation: death certificates from 1983 to 1988 revealed that suicide—and not accidents—was the leading cause of farm-related deaths. In fact, they found that Oklahoma farmers were five times more likely to die from suicide than from farm accidents. In the five-year period studied, 160 farmers ended their lives.[2]

Revelation of this startling information incited officials to action. The governor quickly organized a task force to investigate the cause of the high rural suicide rate. Lenders and lending practices were brought under increasing scrutiny. The existing state agricultural crisis line, AG-LINK, ratcheted up its existing services and expanded the size of its crisis intervention team, making it capable of responding anywhere in the state. The team had made an average of 120 in-home suicide intervention calls each year through 1988. In 1990, just prior to the initiation of my fieldwork, calls had increased to 175.

According to Mrs. Perkins, the AG-LINK coordinator, suicide deaths were highly patterned in terms of method and gender. Self-inflicted gunshot wounds to the head or chest were, by far, the most common method of suicide. And despite the highly publicized death of a farm woman in southwestern Oklahoma who was about to lose her family land (she set herself—and a mound of family memorabilia across which she had lain—on fire), only three suicides had been confirmed for farm women by AG-LINK at the start of this research. At our meeting, Rick emphasized again that this was almost entirely a male phenomenon.

I was propelled to this research by my desire to understand what led farmers to these tragic acts. I wondered what beliefs and feelings motivated men's manifold responses to the farm crisis—responses that more often included withdrawal, silence, and inaction; terrific work efforts and, at times, political activism and even violence. I was told that men committed suicide because of their perceived failures as farmers and as providers for their families. Having been raised in northwestern Oklahoma, I thought this was too simplistic; I knew more than economic failure would be implicated in understanding men's responses to crisis.

I naturally assumed a correlation between the farm crisis and the rising numbers of agricultural fatalities. But working in rural public health, I noticed a disconnection or, at least, a convolution: farmers' destructive responses to the financial crisis were interpreted as resulting from "farm stress" and not from any sort of economic transition. My colleagues, for instance, worked to develop educational materials aimed at helping farmers understand stress and recognize its warning signs. They designed and implemented trainings to teach coping strategies and communication skills. But hardly anyone was talking about the root cause of stress. It seemed that public health and social service practice, undergirded by psychological discourses that are deeply resonant in contemporary America, supported an understanding of men's experiences that was firmly centered on individuals and what they lacked—resiliency, skills, information—and not on the economic crisis in family farming and the changing nature of rural communities. As a result, men's responses to the crisis were interpreted as psychological and individual events.

But as an anthropologist, I wanted to understand the cultural patterns that I witnessed and that led me to believe that men's responses were more than psychological: Why did it appear that it was only men that were responding to the crisis? Why did they respond in the manner in which they did? And why was the emotion of pride used so often to justify men's actions and inactions? I came to believe that pride was the key: it was talked about with such frequency, that it emerged as a central problematic, an issue whose study I thought could illuminate the relationship between men's individual experiences, cultural meaning, and social change.

If pride was at the root of understanding men's actions and inactions, I wanted to know what pride *meant*. Consequently, my principal fieldwork goal became to identify and learn about the cultural discourses that constituted the emotion of pride: What did people associate with pride? What kinds of things did they talk about when they talked about pride? Finally, why were the topics associated with pride so charged that they might make someone commit terrible acts?

In my search for answers to these questions, the farm men and women with whom I worked during the course of my fieldwork led me far beyond an examination of health and mental health to other issues that were central to their self-understanding and that are discussed in this book. These include ideas about gender and kinship, land and farming, agricultural economics and politics, Oklahoma history, American history, and Christianity. They also led me beyond a geographic focus on Oklahoma in at least one sense.

I soon learned that the dynamics of the crisis were not just a local phenomenon, but were part of what I was told was the agricultural sector's evolution toward, and involvement with, prevailing national and international trends in food production. In other words, Oklahoma agriculture was becoming increasingly modernized, though no one called it that.

We generally don't think of "modernization" in America; we think of the adoption of innovations, economic transitions, and restructuring. We assume that modernization ended long ago in the West and that it is, instead, a process that developing countries undergo as their economies industrialize and their societies bureaucratize. Further, the manifestations of modernization and of capitalist development in the third world positively signal to us the readiness of those nations to join the forward, linear march of history—that "they" are becoming like "us."

But my point here is that not all of *us* have become like *us*. While Mitchell (2000) and his colleagues challenge assumptions about the genesis, timing, and space of modernization by examining its multisited creations outside the West, one of its features, they argue, remains universal: modernization's dependence on the systematic subordination and marginalization of social beliefs and practices that are incompatible with it. These cultural forms are assigned a "different and lesser significance" and are "placed in a position outside the unfolding of history" (Mitchell 2000:xiii).

Although the process of modernization in many places in the developing world differs in kind and degree, I argue that most sites share with this particular American context the depreciation of certain beliefs and practices—and the valuation of others—and the experience of social upheaval. For example, during my fieldwork experience in northwestern Oklahoma, I learned that industrial values and the increasing incorporation of technology in the practice of agriculture are having profound consequences for farming communities. I witnessed the disintegration of social relations and communities and a crisis of faith in institutions and values among rural residents. I found "large numbers of persons who [could not] keep up with progress" who have had to have their "expectations of a comfortable life frustrated" (United Nations 1951, quoted in Escobar 1995:3). Willing or not, it seemed these Oklahoma communities were paying a high price for "progress."

This study examines how the modernization, or "restructuring," of the agricultural sector has affected rural Oklahoma communities and, specifically, its impact on the thoughts, feelings, and actions of individuals. I show how these cultural and historical processes of change have challenged not only particular agricultural and social practices but also certain subjectivities: ways of being a farmer and a man and men's identity or sense of self.

I argue for an understanding of farmers' responses to crisis that is not strictly psychological, individual, or idiosyncratic, that farmers' pride and actions are a consequence of a multiplicity of cultural discourses. To illustrate the point, I reconnect pride and men's experiences and practices to their context to demonstrate that any adequate understanding of action—even seemingly individual acts—must recognize cultural constraints and enablers, not just psychological ones. This perspective necessitates a view of emotions as culturally mediated, "embodied thoughts" (Rosaldo 1984:143) that are necessarily evaluative.[3] Such an orientation also acknowledges the lived realities of social, and not just individual, difference and power: that we do not all have the freedom to make and enact the same choices.

Because the patriarchal foundation of northwestern Oklahoma farming communities was being transformed by industrial values, I assert that many men experienced the devaluation of the cultural ideas and practices that supported their subjectivity, specifically the emotion of pride. I argue that as the cultural content of pride was being stripped away, this devalua-tion, in turn, too often led men to actions that were often negative, de-structive, and tragic—actions, nonetheless, that serve as potent critiques of American and global notions of "progress."

When we finished our meeting with Rick, Jackie put her hand on my back and guided me down the front steps of the Department of Agriculture, saying, "Well, it looks like you found yourself a project." But then she stopped, and as I turned to face her, she said more seriously, "You should really do this. I think it's important."

I knew that she was right. And unlike a research project I began and abandoned in midcourse—and for which I have always felt regret—this one I thought was doable. If nothing else, I had learned much about the challenges of fieldwork. This time I knew where and how to begin. After all, I was at home and had extensive personal and professional contacts across the region. What I was less prepared for was the way that my rela-tionship to this place would change as a result of conducting *homework*.[4]

Baby, I know we've got trouble in the fields
When the bankers swarm like locusts
Out there turnin' away our yield
The trains roll by our silos . . . silver in the rain
They leave our pockets full of nothin'
But our dreams and the golden grain
Have you seen the folks in line downtown at the
  station
They're all buyin' their tickets out
And they're talkin' the great depression
Our parents had their hard times . . . fifty years
  ago
When they stood out in these empty fields
In dust as deep as snow
Chorus:
And all this trouble in our fields
If this rain can fall
These wounds can heal
They'll never take our native soil
What if we sell that new John Deere?
And then we'll work these crops with sweat and
  tear
You'll be the mule
I'll be the plow
Come harvest time . . . we'll work it out

There's still a lot of love
Here in these troubled fields
There's a book upon the shelf about those dust
  bowl days
And there's a little bit of you and a little bit of me
In the photos on every page
Our children live in the city
And they rest upon our shoulders
They never want this rain to fall
Or the weather to get colder
Chorus:
And all this trouble in our fields
If this rain can fall
These wounds can heal
They'll never take our native soil
What if we sell that new John Deere?
And then we'll work these crops with sweat and
  tear
You'll be the mule
I'll be the plow
Come harvest time . . . we'll work it out
There's still a lot of love
Here in these troubled fields

—"Trouble in the Fields," Nanci Griffith and Rick West

ONE

The Invitation to Die

WHILE I WAS away at college in Vermont, I first became aware of the economic crisis affecting American family farmers through the media. During the early eighties, stories were emerging, mostly from the plains states, about the numerous foreclosures on farms, the displacement of families, and the psychological effects of the loss of land on farmers—land that had often been in families for generations. This latter aspect of the crisis seemed to garner the most attention: newspaper articles, television news coverage, and even feature films focused on the growing problem of suicide among financially burdened farm families.[1]

As a young person trying to distance myself from my Oklahoma upbringing, these stories, initially, were somewhat emotionally removed.

They would not remain that way for long. In the late 1970s, my parents had sold the family business at a high point in Oklahoma's farm and oil economies. Enid, my hometown and the agricultural and cultural center of northwestern Oklahoma—the geographic area on which this study is based—was experiencing dramatic economic growth. The wealth of this small city of 50,000 was evident: there was liberal spending, a significant increase in commercial and residential construction, a population boom, and the infectious excitement that came with the seemingly boundless potential for opportunity and growth.

With the simultaneous collapse of the agricultural and petroleum industries in northwestern Oklahoma in 1982 and 1983, things changed; grim stories of ruin, once remote were suddenly at hand and commonplace. My parents had to scramble for a new way to make a living as the buyers of our family business went bankrupt. My visits home from college turned into solemn occasions—not only because of our own financial situation, but also because of the economic crisis gripping our community, affecting friends and neighbors.

At the time, the cause of the economic crisis was not clear to me; I thought we were experiencing a natural, rhythmic economic downturn—albeit a severe one. Since then, I have realized that the situation was significantly more complex. Northwestern Oklahoma, like almost all agricultural areas of the United States, was being affected by particular national and international economic and political phenomena that converged in the early 1980s.

During the early 1970s, the future of the U.S. agricultural sector appeared promising. There was evidence that overseas markets were ready to open and be eager consumers of American grain. In addition, the Soviet Union had been, and continued to be, a generous buyer of U.S. wheat. The realized growth in demand, combined with the low value of the dollar and tax and farm policies that encouraged investment in agriculture, helped to position the United States to be the "bargain supplier in world export markets" (Barnett 2003:164). One result of this confluence of forces was that the price of wheat was pushed to new highs. Seeking to take advantage of the auspicious environment, many farmers, understandably, sought to expand their farm operations. These expansions, however, were financed primarily through debt capital.

As farmers accumulated debt, the price of farmland began to rise dramatically. This was due to the investment incentives already mentioned, an inflationary U.S. economy, and the high returns available in agriculture, making land an increasingly valuable commodity and a hedge against inflation. But as international demand for grain was met and wheat prices began to level off and eventually fall, many producers found that they were unable to continue to pay for their high-priced land with their agricultural earnings. Farmers were placed in an ironic position: even though their "paper" wealth was expanding, due to the continuing rise in land valuations, they were having cash-flow difficulties. This problem was largely solved through the refinancing of debt, a mechanism lenders believed was still viable because of the ongoing increase in the price of land, their principal form of collateral.

To put the brakes on the inflationary economy, which had reached 11.3 percent in 1979, the Federal Reserve sought to constrain the growth of the money supply. Their action succeeded in increasing the value of the dollar, but it also negatively influenced exports, since American products were now more expensive for foreign buyers. In addition, the increasing federal budget deficits of the 1980s—due to Reagan's supply-side economic policies—significantly raised interest rates, a reality that had a dramatic effect on the highly leveraged agricultural sector.

The combination of high interest rates and reduced farm incomes resulted in a sharp devaluation of agricultural assets, principally land. As a result, many farmers found that their lenders—because their collateral was quickly losing value—were not willing to refinance producers' debts when they were no longer able to make payments. For many farm families, then, the price of land dropped to a level that was insufficient to secure the outstanding loans that they had obtained at high interest rates in better economic times (Rosenblatt 1990:3; Barlett 1987:30). This turn of events was ruinous for many farm families and the towns in which they lived. Farm foreclosures, bank insolvencies, the loss of local businesses, population migration, and even school and hospital closures were the new reality with which local communities had to contend.

The farm crisis of the 1980s, however, cannot be understood in historical isolation, but must be read as a particularly dramatic chapter in the evolutionary story of American agriculture. Rapid modernization and technological advancements have profoundly influenced the practice of agriculture throughout the twentieth century (Fitzgerald 2003). Moreover, these factors were not, and are not, isolated to the United States, but are processes at work within the context of a world market economy. Many anthropologists have documented the cultural dynamics of change associated with farm mechanization and technological innovation, variously affecting cultures across the globe (Gupta 1998; Collier 1997; Delaney 1991; Rogers 1991; Harding 1984).

Cognizant of these processes of change, many researchers—instead of using the term "farm crisis," which denotes an isolated space in time—now speak of the ongoing restructuring of the American agricultural sector. It is important to note that this restructuring is of a particular kind, characterized by increasing industrialization, or the consolidation and vertical integration of production—the same kind of economic transition that has considerably influenced the course of other professions (e.g., fishing, publishing, and medicine). Indeed, American agriculture is being transformed from a "way of life," with a focus on family-owned and operated farms, into a business.

The restructuring of the agricultural sector has had, and continues to have, a profound effect on rural cultures and peoples. In this book, I examine the effect that this restructuring crisis[2] has had on rural communities and focus specifically on the personal consequences of dramatic cultural change for rural individuals—particularly men.

I graduated from college and entered a doctoral program in anthropology at a California university. Upon completion of my course work and comprehensive exams, I decided to take some time off from graduate school to seek employment that would complement my studies. As a medical anthropologist, I wanted to acquire applied skills in public health that I believed would be of value in future career endeavors. Consequently, beginning in 1989 I worked as an education coordinator for a rural public health agency, Rural Health Projects, Inc. (RHP). RHP, located in Enid, began as an Area Health Education Center, a division of the U.S. Public Health Service. Congress had established these regional centers in the early 1970s in response to a Carnegie Commission report that indicated that the health of Americans was being significantly affected by the shortage of health professionals in rural areas and in inner cities. The Area Health Education Centers were specifically mandated to work in collaboration with medical schools to address the maldistribution of healthcare providers by encouraging their placement in government-designated "Health Professionals Shortage Areas" and by maintaining the skill level of practitioners already working in these areas, through continuing education.

While at RHP, I was responsible for assessing community health needs through methods that primarily included key informant interviews and focus group research. Based on my findings, I would design educational programs to increase the capacity of rural health professionals to respond to locally identified needs. In addition, I was heavily involved in grassroots

organizing, encouraging individuals in local communities to collabora-
tively address shared health concerns. My research began informally, then,
in 1989 as I became reacquainted with the land, history and people of my
home territory while working for RHP.

I typically traveled for periods of three to four days per week to rural
communities to gather information and to participate in regional working
groups that targeted specific health issues, such as farm-related stress,
teenage pregnancy and child abuse. Through that experience, it became
clear to me that it was difficult for participants to discuss particular health
concerns without also talking about broader societal processes, such as
poverty, marginalization, racism, education, and gender roles. What I
learned from my colleagues, in short, was that inequality, in its various
forms, has a significant and adverse effect on health.[3]

Chief among the social issues discussed by my colleagues was the regional decline in the agricultural sector, which, in turn, initiated a downward spiral of cause and effect that compromised the future of rural communities. My peers talked about families barely being able to make a living
from wheat and cattle farming. As a result, they told me, many chose to
move to urban centers, leaving a smaller customer base for local businesses. This, in turn, caused numerous businesses to close, producing the
boarded-up Main Streets that characterize so many rural Oklahoma
towns. Fewer and fewer economic options fueled a continuing population
drain, making it much more difficult for local communities to support an-
chor social and health services, such as hospitals and, indeed, many have
closed or are closing.

These economic realities had other social consequences. I discovered
individuals struggling to make sense of their rapidly changing communities, struggling to make sense of their place in a world whose fundamental
ground rules, they believed, had changed. They bemoaned the loss of
community and the lack of career options for their young people and their
inevitable departure. Most insidious of all, they told me of the increasing
competitiveness of neighbors.

These social phenomena had individual consequences for health. In
hushed whispers they spoke of neighbors' problems with alcoholism, domestic violence, child abuse, divorce, suicide, and other issues that are, in
public health, traditional indicators of stress in communities—and that researchers have correlated with increased stress levels among financially
burdened farm families (Jurich and Russell 1987).

Statistical data bore witness to communities in obvious economic
straits. A statewide survey of farm families conducted in 1990 indicated
that 20.8 percent of Oklahoma's farm and ranch families had debt-to-asset

ratios of 40 percent or more (Oklahoma State University Cooperative Extension Service 1991:1). That is, for every dollar of assets, these farmers owed at least forty cents, indicating serious economic difficulty. While real interest rates on the majority of farm debt was still near record highs during my fieldwork, land values remained 40 percent below the market peak reached in 1982. As a result, record bankruptcy rates were documented in Oklahoma just prior to the start of my research—14,116 individuals in 1989 alone (Oklahoma State University Cooperative Extension Service 1991:1).

Of particular note was the suicide rate among farmers—three times that of the general population. From 1983 through 1989, one hundred sixty Oklahoma farmers committed suicide. In just a one-year period, 1988 to 1989, forty-three suicides were confirmed (Oklahoma State University Cooperative Extension Service 1991:Executive Summary). But there was a particular pattern to these suicides: men committed the vast majority. In fact, 84 percent of rural suicides were committed by men as compared to 77 percent for the rest of the state (Oklahoma State Department of Health 1991).

These statistics struck me as curious. I knew that financial stress strikes entire families. I also knew that a case could be made for women being as strongly affected as men by actual or potential farm loss and severe indebtedness; women were, by and large, responsible for the fiscal administration of the household and, often times, for agricultural bookkeeping. They, after all, would be faced with the day-to-day economic decision making that comes with maintaining a family and running a household; they, too, would know when the farm financial numbers were "not adding up."

But when I mentioned these statistics to my colleagues who work in the health and social service professions in northwestern Oklahoma, no one seemed surprised by the gender patterning. One colleague said, "Well of course, that's natural, Eric. A man would rather die than admit he's failed. Pride is all it is." This was a common sentiment.

Transcripts of public hearings held around the state regarding farm-related deaths, together with preliminary interviews I conducted, indicated that male farm suicides were anything but "natural." One farm woman described farmers' responses to crisis, and demonstrated their link to culture, in a prepared speech at one of the public hearings held in northwestern Oklahoma:

> Living in a society where self-esteem is closely related to personal wealth and status, the family in distress must cope with a loss of stability and economic un-

certainty, as well as a sense of failure and personal isolation. Most important, the community and personal support that could be helpful is often lacking during this period of crisis. Now comes the invitation to die from others in our lives. A spouse strikes out because of heavy stress, the equipment dealer repossesses the equipment, and the land bank forecloses on the land. The lenders announce publicly in the newspapers that you have failed. Embarrassed and humiliated you withdraw from all community activities and go into hiding. The loss of self-esteem, self-dignity, and self-worth is intensified by community rejection as a poor manager or business failure. Where can they go and what can they do? There doesn't seem at this point but one thing to do: suicide.

This informant's comments reveal that crisis may initiate culturally patterned responses that too often end in tragedy. Specifically, they describe a society that values personal wealth and status as indicators of success; success denotes cultural expectations and evaluation. They disclose that failure incurs social censure, manifested through community isolation, and that the evaluation of failure is often shared by those who have failed and serves as motivation for self-imposed isolation. This failure produces embarrassment and humiliation—culturally accepted emotions for one who does not live up to social expectations of economic success. This informant's comments betray the individualism of the region by attributing failure to individuals (calling someone a poor manager, for instance) and not to macroeconomic and political factors. Finally, they reveal that the identification and censure of individuals—"the invitation to die"—is both a social process and an individual experience, productive of the loss of self-esteem, dignity, and self-worth, providing the motivation for suicide.

Implicated, then, in suicide and the experience of severe financial distress are deeply gender-linked cultural factors. They include among farm men: (1) the equation of masculinity and sense of self-worth with the financial "bottom line," (2) the social support available—or unavailable—both within and outside of the family setting and the cultural proscription against most emotional expressions, (3) an honor system in which the "capacity to demonstrate autonomy through making and enacting choices" (Errington 1990:639)—including economic ones—is central, and (4) an intense attachment to land, which is tied to notions of gender and kinship.

As I prepared to conduct research and reviewed the anthropological literature on American farming culture and, specifically, the farm crisis of the eighties, it struck me as curious that none of the literature dealt with the aspect of the crisis that has received most attention: farm or land loss and its effects on families. The literature generally interprets the farm crisis

and the ongoing restructuring of American agriculture as simply an economic phenomenon with psychological consequences.

Why the silence on the part of anthropologists concerning this particular aspect of the crisis? Since the effect of the farm crisis can be understood
as individual or psychological, did anthropologists believe that this topic
lay outside their area of expertise? Since anthropology relies so heavily on
the elicitation of informants' beliefs for data and interpretation, does the
silence have anything to do with the discipline's construction of its subjects
as rational—and not emotional—beings? I wondered whether ideas about
gender had something to do with the paucity of data on the topic. Did the
traditionally male-dominated academic establishment value studies focused on emotions as less rigorous or as less important? Or is it simply because farm researchers did not expect to find emotions among men?

I was drawn to these "psychological consequences" of the crisis; I
thought that their examination would provide an important means to understand the cultural notions informing American farm life and the dynamics of agricultural restructuring. I also believed that I could come to
better grasp the theoretical problematic that had always intrigued me and
that is a core issue of the social sciences: the relationship between individuals, their agency, and social structures. I thought that by melding accounts of social practices and individual experiences concerning the
"force" (R. Rosaldo 1989) of emotions, that I could come to an understanding of complicated actors and their actions: how individual histories
influence perceptions of cultural options and social action, how action is
motivated, and the role that intensity of feeling plays in human action. In
addition, I believed that I could utilize emotions as a vehicle to comprehend the cultural dynamics operating in different spheres: the individual,
the local, and the global.

Thus, in this book I argue that responses to the farm crisis are not
strictly individual and idiosyncratic, but rather that cultural discourses of
meaning position actors in structures of power and condition their actions
and subjectivity—or sense of self—and its expressions. In analyzing male
farmers' responses to severe financial indebtedness and their ideas about
masculinity and male roles, then, I seek to illuminate how gender identity
is mediated by emotions and translated into social action and inaction.

I conducted fieldwork from October 1991 to June 1994, primarily in the
area of northwestern Oklahoma called the Cherokee Strip.[4] This portion

of the state, which encompasses several large counties, had been part of the Cherokee Nation prior to non-native settlement in 1893. The U.S. government, feeling the pressure to open the land to new settlers, forced the Cherokees, via threats based on alleged nonobservance of existing treaties, to sell the land to them at prices substantially below market value.

Northwestern Oklahoma was opened to settlement at a time when the American frontier was closing. The region represented the last great opportunity for a new life and a new start for many. Oklahoma became a refuge for misfits, the restless, or those that "openly rejected the rigid structures of the South, or who sought escape from the all-too-settled and urbanizing life in the middle, northern, and eastern regions of the country" (Stein 1987a:189).

Like its settlers, the area's beginnings were unusual. Northwestern Oklahoma was opened to non-native settlement in 1893 via a land run. Settlers wishing to claim 160 acres of free land lined up at the Kansas border and at a line drawn north of Oklahoma City. On the appointed day, 16 September 1893, a shot was fired that signaled the beginning of a great race. On horses, carriages, and anything else that would carry them, men and women vied to stake a claim. The settlement pattern, square mile blocks each containing four 160-acre sections, was prearranged by government officials.

Even so, northwestern Oklahoma communities reflect the spatial arrangement that characterizes much of the Midwest and that, Arensberg (1955:1154) asserts, is distinctively American. The arrangement is known as *Einselhofsiedlung* or the "open-country neighborhood":

> Yankees went to the frontier in wagon trains, to planned villages, and Southerners, some of them, to plantations and country seats cut from the virgin woods, but the Midlanders did neither. They went singly, family by family into the lands they cleared simply by accretion of farms into "neighborhoods." Their first communities were mere crossroads where scattered neighbors met. Their schools and churches and stores, like their camp meetings and fairs, were set haphazardly in the open country or where roads met, with no ordered clustering and no fixed membership.
>
> (Arensberg 1955:1154–1155)

The "open-country neighborhood" of the Middle West did not spontaneously arise, however. Many of the settlers to the Midwest came from the middle Atlantic region, which, in turn, received its immigrants from the places in Europe that practiced *Einselhof* dispersal of individual farms.

When they left the mid-Atlantic states and went to the Middle West they brought this settlement pattern with them. Stein argues, then, this use of space was familiar to these migrants and thus were "*preadapted* to settle and inhabit the new environment in the way that they did" (1987a:192).

In addition to spatial arrangement, Arensberg asserts that the culture settlers brought with them to the region was also distinctively American. The Midwest and the Appalachian Frontier were "the first American regions to stand clear, to rise out of mixture and to shape new and free conditions" (1955:1153). Unlike the communities one finds in New England and the South (when Arensberg was writing), whose majority members share some common ethnic stock as "Yankees" or as "Southerners" of English ancestry, the Midwesterners have no such common bond—as the region is a direct heir to "that seedbed of mixture that parented them," (1955:1153) the Middle Colonies, or the mid-Atlantic states. The variety of peoples that came to the mid-Atlantic states and went on to settle much of the Middle West—Scots, Welsh, Irish, Germans, Scandinavians, Slavs, Latins, and English—forged a truly "American" identity out of this diversity.

This, in essence, also describes the experience of northwestern Oklahoma. The race for land in the Cherokee Strip mitigated the formation of ethnic communities of the kind described, for instance, by Salamon (1992) in her comparative study of German and "Yankee" farming communities in Illinois. This reality may be the basis of Stein's assertion that "a sense of Oklahoman, regional and American identity overrides ethnonational origins and private traditions" (Stein 1987a:193), since opportunities for the reinforcement of traditional cultural practices were less available and less widely sanctioned in the communities of new settlers.

In addition to settlement pattern, the settlers to the region may also have been preadapted to the rough-and-tumble world of the Oklahoma frontier in another sense:

> Those who left their European homelands in search of freedom—or, those who left their parental settlements in the American East or North in the quest of independence—were already rebels against their native traditions and were necessarily dissatisfied with and somewhat separate from it.
>
> (Stein 1987a:192)

The hardiness of rebels is perhaps what was required since settlers to Oklahoma were faced with many obstacles. The condition of the land at the time of non-native settlement by no means ensured the success of claimants. In fact, most failed. Historically, the region also had to contend with the challenges of the Great Depression and the ecological disaster known as the Dust Bowl. The stories of ancestors' survival, despite the nearly impossible conditions, are important to my informants and critical to apprehending the cultural dynamics of the current crisis, a topic which I discuss in a later chapter.

Northwestern Oklahoma, however, had some assets early on that helped the region's residents not only survive, but also thrive. The area's principal towns—Enid, Woodward, and Ponca City—share three characteristics that helped them to achieve sustained development: (1) an early transportation infrastructure that facilitated the growth of, and access to, services and markets, (2) effective local leadership, and (3) some measure of economic diversification (Turner and Gailey 1998:117–118). These factors, combined with historical luck, contributed to the development of substantial human resources and material infrastructures that poised northwestern Oklahoma for continued growth and prosperity—a significantly different experience from that of the rest of the state. This "head start" continues to make a difference today: the counties with the highest per capita incomes are found in this quadrant of the state. Visible poverty is not present to the same degree, nor is it of the same kind, that one finds in disturbingly high rates in other areas of Oklahoma. An alternative explanation for the seeming prosperity of the region in light of the farm crisis is that individuals who have lost their farms or who are unemployed have little choice but to leave the area because of the lack of other economic opportunities.

Northwestern Oklahoma is composed of rolling plains that, to the unaccustomed eye, may seem flat. The land is also more arid and almost treeless when compared to the European and eastern American communities described by Arensberg (1955) in his now classic article. Finally, the territory is very sparsely populated. The entire northwestern quarter of the state holds only 10 percent of Oklahoma's population of more than three million. Some counties in the area have population densities of less than four people per square mile—considered "frontier" counties by federal statistical standards. Besides Enid (with a population of 50,000), Ponca City (26,000), and Woodward (12,000), only a small number of communities have populations of a few thousand and the majority contains only several hundred, if that.

Northwestern Oklahoma was an ideal location for research focused on the farm crisis. The land is well suited to wheat production, which, in turn, is a perfect temporal complement to cattle production. As a result, the northwest contains the largest proportion of full-time farmers in the state and has been, arguably, most affected by agricultural modernization and restructuring.

The high concentration of farmers is also most likely the reason that the conflict between the dominant values of production-focused industrial agriculture and those of family-centered farming is most apparent in this region. Comprehending the nature of this tension—which plays a pivotal role in this study—is critical to understanding the farm crisis, the cultural and historical processes of power driving rural social change, and the personal effects of agricultural modernization that is at the heart of this book.

In the context of the rapid social transformation that characterized my fieldwork site, resistance to change and to hegemonic industrial discourses existed in the form of the American Agriculture Movement, an activist organization that opposes industrial interests and advocates policies that support family farmers. The philosophical and practical divergences of industrial and family-centered agricultural practices can be most clearly delineated by examining farm management styles (Bennett 1982), or how and why individuals farm.

Salamon's (1992) work is illustrative. She has outlined two management styles for farmers—yeoman and entrepreneur—which she found correlated with ethnic background in her Illinois sample of German and "Yankee" communities. For instance, yeoman farmers of German descent hold a more family-centered approach to agriculture. They view land as a sacred trust that they honor by maintaining land within the family through time. They are therefore concerned with ensuring that every generation produce at least one farmer. They employ several strategies to realize their goal of family continuity: farmers avoid financial risk whenever possible, they maximize family labor on the farm, and they limit the operation's growth to match the labor capacity of kin. As a result, yeoman farms tend to be smaller than average.

Yankee entrepreneurs, on the other hand, take a more industrial approach to agriculture. They view land as a commodity and farming as a business whose expansion can serve to increase family wealth. Their goal is to effectively and efficiently manage their farm operations to ensure short-term profits. Again, they employ diverse strategies to achieve this goal, including farming a combination of owned and rented land to optimally use farm equipment, expanding their operations (limited only by available capital) and when necessary incurring substantial financial debt. Their farms, consequently, tend to be larger than the average (Salamon 1992:93).

Barlett (1993) also developed two management models—cautious and ambitious—that are similar to those styles described by Salamon, but vary in their criteria because of the different cultural context of her fieldwork. Both anthropologists make compelling cases for their models, and they both have high analytic and heuristic value because they are empirically grounded in the cultural contexts in which they were elaborated. But because these farm management models represent ideal types, they offered limited guidance in understanding the motivations of the men and women who participated in this research and their consequent actions and inactions in the face of crisis—a central goal of this study. My fieldwork revealed, for example, that few informants would fit easily in either discursive category. The majority merged ideas from both and even other discourses.[5] A notable exception was the members of the AAM, whose espousal of family-centered agricultural practices and values clearly placed them in Salamon's "yeoman" category.

So how can we account for the diversity of farm goals, strategies, and actions in a way that captures the motivations of the men and women who participated in this study? I argue that the answer lies in a conceptualization of culture that is able to account for the complexity of social agency and thus offer an alternative to the dualistic and static frameworks that have dominated much of the anthropological literature on American agriculture since the 1980s.

My incipient awareness of disparate farm management styles and diverse agricultural practices challenged my preconceived notion of farmers as a homogenous group, which, in turn, gave me cause to rethink my beliefs regarding the nature of rural communities. I realized that growing up, I thought of the rural communities surrounding my hometown as social wholes—as unified, culturally homogenous, and socially supportive. As my research progressed, I suspected this conceit was flawed, and realized that the notion of culture, as traditionally conceived in anthropology, would not serve as a useful theoretical construct.

In fact, the idea of culture as a unified entity being shared by all members of a society is being increasingly criticized across the social sciences. Culture, within anthropology, has traditionally implied a "certain coherence, uniformity, and timelessness in the meaning system of a given group" (Abu-Lughod and Lutz 1990:9; see also Stromberg 1986 and Yanagisako and Delaney 1995). The study of complex societies by anthropologists, however, has illuminated an alternative reality, a reality that acknowledges divisions within societies that contribute to the production of diverging ideas, social practices, and personhoods. These divisions include, but are not limited to race, class, ethnicity, gender, and sexuality. So if culture is shared, "we must now always ask 'By whom?' and 'In what ways?' and 'Under what conditions?'" (Dirks, Eley, and Ortner 1994:3)

Power, critics argue, is productive of divisive cultural boundaries. That culture is grounded in unequal relationships—due to power—is an assumption that precedes cultural examination by poststructuralist theorists. Yanagisako and Collier (1987:39), for instance, assert that this assumption is based on the reality of social life: "By most definitions, a society is a system of social relationships and values. Values entail evaluation. Consequently, a society is a system of social relationships in which all things and actions are not equal."

They go on to explain the advantages of such an assumption to an understanding of culture:

> The premise that all societies are systems of inequality forces us to specify what we mean by inequality in each particular case. Instead of asking how "natural" differences acquire cultural meanings and social consequences, a presumption of inequality forces us to ask why some attributes and characteristics of people are culturally recognized and differently evaluated when others are not. This requires us to begin any analysis by asking, what are a society's cultural values? And what social processes organize the distribution of prestige, power and privilege?
>
> (1987:40)

The assumption of culture as grounded in unequal relationships signifies a shift in cultural analysis that is significant for several reasons: (1) power comes to the fore as a primary analytic tool, (2) no longer are institutions and other traditional structures of society seen as the sole arbiters of cultural meaning, but rather (3) everyday life becomes the focus of analysis, as individuals, social groups, institutions, and the state contest the legitimacy of different forms of knowledge, interpretation, and authority.

I argue that as an alternative to the traditional notion of culture as being shared by all members of a society, a *processual* view of culture (Stewart 1996; Dudley 2003; and Yanagisako 2002) facilitates a better apprehension of every day life, including the complexity of social agency and the diversity of actions of individuals. A processual view entails an understanding of culture as being *"constituted in use* and therefore likely to be tense, contra-dictory, dialectical, dialogic, texted, textured, both practical and imaginary, and in-filled with desire" (Stewart 1996:5; emphasis added). Such an ap-proach necessitates a focus, naturally, on *culture in use,* or the practices and actions of individuals in their daily lives. In essence, individuals *take* from culture, drawing from various discourses[6] of meaning to formulate action and a sense of self, including emotions, and they also *make* culture through their actions and modes of being, "always sustain[ing] and sometimes transform[ing] the very structures that made them" (Dirks, Eley and Ort-ner 1994:12).

My point is that a processual view of culture requires that we examine the relationship between social action and individual subjectivity or sense of self—a project that is a principal theme of Dudley's (2000, 2003) work with northern Plains farmers.

Dudley's study makes several important contributions. First, she high-lights an important reality: that despite a diversity of goals and strategies—whether yeoman or entrepreneur—all farmers are well aware that they are producing for capitalist markets. Arguing against much of the existing lit-erature, Dudley asserts that it would be misguided to assume, "that the re-lentless decline in farm numbers—and specifically, the economic practices which put farms at risk in the 1980s—is evidence of the encroachment of 'industrial' or 'capitalist' values and the erosion of an otherwise stable, 're-publican' or 'agrarian' cultural system" (Dudley 2003:178). Rather than seeing divergent agricultural practices as emanating from distinct cultural forms, Dudley understands that this diversity stems from the same cultural system: the logic of capitalism. The linchpin of this capitalist culture, she posits, is a distinctive notion of self, the "entrepreneurial self," which holds that an individual is "personally accountable for the consequences of eco-nomic risk-taking" (Dudley 2003:176).

Dudley thus offers a theoretical framework that is dialectical and dy-namic, rather than dualistic and static: diverse management goals, strate-gies, and motivations are understood as being both formed by, and forma-tive of, capitalism and its cultural logic. This is true, she argues, since all farmers, ultimately, share the same interests: individual achievement and profit maximization. With the notion of the "entrepreneurial self," Dudley further demonstrates that a distinctive conception of self, or identity, emerges through capitalist practices at the local level, arising from a cul-tural commitment to economic growth which instantiates a particular communal morality—a morality that promotes social and capitalist prac-tices that radically individualize and isolate farm families through competition. It is a form of community that reproduces capitalist cultural logic and "*endorses* the very actions that give rise to the disorder its members experience" (Dudley 2003:180).

For Dudley this entrepreneurial self is based primarily upon informants' economic understanding and behavior, that it is both formed by, and formative of, a "capitalist logic."[7] However, because the diverse actions of farmers seem self-evident in light of this capitalist logic, I argue that the role of other cultural influences is closed off as a possible avenue for understanding the motivations of farmers and their subsequent actions and inactions. Could it be that other spheres of social life, beyond the economic realm, also influence economic motivations and behavior and explain the diversity of capitalist and agricultural practices?

Yanagisako (2002) believes this is indeed the case. That this is a question at all, she argues, is due to the historical strictures of Western social theory—primarily the work of Weber—that have supported a dichotomous understanding of social action, making a clear distinction between *economic* action and *other kinds* of action. Economic action has been traditionally understood as being motivated by rationality and calculation: objective and utilitarian interests in the pursuit of profit and capital accumulation. As a result, she argues, affective or emotional motivations have been relegated to explaining "other" social action, such as that in familial and religious realms. "The instrumental/affective dichotomy," Yanagisako (2002:9) asserts, "has facilitated the acceptance of the concept of *interest* as a motivating force for economic action and the failure to recognize its emotional component." This instrumental/affective binary, she continues, has served as an obstacle to viewing all social action "as constituted by both rational calculation and by sentiments and desires: in other words, as cultural practices" (Yanagisako 2002:21).

Yanagisako's ethnography on northern Italian family silk firms serves as an exemplar of a more holistic approach to understanding social agency. Her work elucidates the important roles that sentiments and cultural notions of gender and kinship play in motivating economic behavior and its effects on family firms: their management, expansion, succession practices, division, and diversification. Her work not only helps us to understand motivations as being composed of both a rational calculus *and* emotions and desires, but it also facilitates our apprehension of the diversity of agricultural practices and management styles which were recognized and outlined by Salamon and Barlett. Since farmers may not share the same personal trajectories, power, or status within a hierarchical social structure and, ultimately, may not share the same interests, they are likely to draw from diverse discourses—not only from the economic sphere, but also from the arenas of politics, religion, emotions, gender, and kinship—in "formulating meaningful social action" (Yanagisako 2002:194, n. 20), resulting in diverse cultural or agricultural practices. It is through these practices, "including what are conventionally construed as 'business' practices" (Yanagisako 2002:5), or agricultural practices, that farmers' identities and orientations are constituted.

This means that we cannot speak of capitalists—or farmers—as a "homogeneous, undifferentiated group" (Yanagisako 2002:8). So instead of one capitalist logic as posited by Dudley, there in fact can be several "capitalist logics" informing diverse practices and subjectivities: capitalism is a "complex and uneven historical process that entails heterogeneous capitalist practices shaped by diverse meanings, sentiments, and representations" (Yanagisako 2002:7), practices that coexist within the same geopolitical space.

If Yanagisako's formulations help us to understand the motivations and the diverse agricultural practices of the men and women who participated in this study, they also lead us to ask other questions: Why are some capitalist or agricultural practices and identities more valued than others? And how is it that certain agricultural practices and ways of being a farmer come to represent the interest and identities of farmers as a whole?

The answers to these questions, I argue, lie in understanding the discursive matrix in which agricultural relations are embedded and particular agricultural practices, subjectivities, and the social structures they support are contested—a project that is the crux of this book.

This study, then, goes beyond the traditional focus on economic and political interpretations of the farm financial crisis and its enduring aftermath, to incorporate gender, emotions, and other cultural discourses to understand the motivations and experiences of farm men and their responses to crisis. By viewing individual subjectivity and cultural practices in relationship to one another within a system of inequality, I demonstrate how the intensification of capitalism in the agricultural sector and cultural and historical processes of power impelling rural social change not only devalue certain agricultural goals, strategies, and actions—specifically family-centered approaches to farming—but also devalue particular subjectivities or ways of being a farmer and a man, too often leading to tragedy. In doing so, this study elucidates the process by which gender, emotions, and other historically situated discourses beyond the economic sphere serve not only as powerful forces of production, but also of destruction.

Doing Fieldwork: An Anthropologist at Home

In her article, "Dissolution and Reconstitution of Self: Implications for Anthropological Epistemology," Kondo (1986) discusses the perspectives that anthropologists bring to the field and their implications for research and interpretation. She tells us that anthropologists' perspectives can vary according to their position with respect to the society they are studying. "Position" here refers to how removed or alien the anthropologist may find the culture of her research site—in other words, the status of the researcher as a relative insider or outsider. Kondo argues that, inevitably, this relative positioning affects the interpretation of the society that is produced. I have no doubt that my fieldwork experience and its result—this study—were influenced by my professional and personal roles: as staff member of RHP and as someone who was raised in northwestern Oklahoma.

As I mentioned in the introduction, just prior to the initiation of my fieldwork, RHP negotiated a contract with the OSU Cooperative Extension Service to address the mental health needs of rural families. The goal of the resulting project, "Helping Hands," was to improve the emotional well-being of Oklahoma farm and rural families by maximizing the accessibility and acceptability of mental health services. A secondary goal of the project was to elicit support from the community and the mental health system for proactive and preventive mental health services, earlier use of supportive services, and less reliance on crisis intervention. Professionally, I dedicated twenty hours per week to "Helping Hands." For the rest of my time, I was granted unpaid leave to conduct independent research. I was fortunate in that I was able to draw from my experience and insights with "Helping Hands" for this study. Both projects, at their core, are concerned with understanding the personal experiences of those affected by the farm crisis.

As part of our work with "Helping Hands," another investigator (Lucia Rojas-Smith) and I conducted *focus group research*. We were interested in eliciting ideas about gender, farming, the meaning of land, and other cultural issues to better understand the beliefs that underlie health behavior and help seeking. We conducted seven focus groups with diverse populations: three were organized with farm men, two with farm women, and one each with high school youth and older rural residents. Typically, the meetings consisted of eight to twelve individuals, recruited by county-based OSU Cooperative Extension agents. Dr. Rojas-Smith and I served as moderators. Together we constructed a short list of questions to be addressed by the group. The format, however, was open-ended and we encouraged an in-depth exchange among participants. The meetings lasted approximately 90 minutes. The resulting data were subjected to thematic and content analysis in the interpretive tradition.

Independently, and in the early stages of this research, I used my professional contacts through RHP and "Helping Hands" to conduct both *informal and more formal open-ended interviews* with therapists and social service and health professionals who work with farm families. I used these interview sessions as opportunities to familiarize myself with the effects of farm financial stress on families and to learn about their patterns of health seeking. These sessions were rich sources of information concerning the relationship of the crisis to other spheres of social life, including ideas about gender and the dynamics of rural communities. Interviews with colleagues, who themselves had farm backgrounds, were especially helpful.

It was interesting to note that these sessions often harbored warnings. I was told that conducting research on men and the rural crisis would be problematic because farmers: (1) don't talk very much and (2) would probably not want to talk about topics as personal as their financial status, ideas about gender, their experiences of financial stress, and land loss. These things did not turn out to be true. I believe this is due, in part, to the appropriateness of the *collection of life stories* as a research method for this population.

The collection of individual life stories constituted the primary source of data for this study. I gathered twenty-six life stories, with an equal number of interviews conducted with men and women. Since the collection of a single couple's life histories usually required three to four visits (each approximately 1.5 to 2 hours in length), there was time for some rapport to develop and some degree of trust to be established between my informants and me. As my research progressed, I learned to ask my less relevant questions during our introductory sessions. As our mutual comfort level increased, areas of direct concern for this study could be explored more openly. Also, I found that this was a somewhat empowering research method. Participants were largely responsible for the order and, to some degree, the subjects we would discuss. I believe this eased farm couples' anxieties, increased their sense of control and trust in me, and created an experience that was usually a pleasure for all involved.

My principal interests in collecting life stories were to understand cultural meanings associated with manhood and farming; the manner and the domains in which ideas about manhood and farming were discussed, negotiated and acted upon; and the historical consciousness that influenced modern notions of the same (Yanagisako and Collier 1987:38–48).[8] For instance, I wanted to learn about the relationship between cultural ideals about "man" and "farmer" and the challenges the ongoing restructuring of the agricultural sector posed to both. Furthermore, I wanted to understand how farmers explained deviations from cultural ideals and if the disjunction between ideals and experience gave rise to other ways of being male (Ginsburg 1989). Within these life stories, I looked for themes that illustrated the role of emotions in conditioning social understanding and action. For example, I examined what these stories revealed about intentions, motivations, social relations, and actions.

A premise of this study is that "man" and "woman" are not natural categories in the sense that biological "facts" do not exist outside of a cultural frame of reference in which they become meaningful. Instead of assuming difference, I chose to question difference: its nature, basis, and meaning within societies to "denaturalize" categories and elucidate the workings of power. For me, this meant trying to understand the diversity encompassed within the category "man." Which ways of being male are valued? Which are devalued? And how are these meanings constituted through social practice?

Gladwin (1989) has demonstrated that a by-product of capitalist development in the farm sector is increasing differentiation of class and land tenure patterns. I anticipated that the information gathered from life stories would differ depending on the economic positions and life experiences of the farmers I interviewed. Thus, to get an idea of the diversity encompassed within the concept of manhood, I gathered life histories from a diverse group of farm men and women: (1) financially burdened and dislocated farmers, (2) financially successful farmers, and (3) farmers who have become political activists with the American Agriculture Movement.

I used the following criteria in selecting informants for this study: (1) Informants had to be full-time family farmers. Recognizing that since 1980 an increasing proportion of operations are rented rather than owned, I adopted Nikolitch's management and labor–based definition of a family farm: "a primary agricultural business in which the operator is a risk-taking manager, who with his family does most of the farm work and performs most of the managerial activities" (Nikolitch 1969:531). (2) Informants needed to be wheat and cattle operators, the predominant pattern in northwestern Oklahoma. (3) Informants had to manage farms that fell in the mid-size range for this area, which is approximately 500 to 1,000 acres. (4) Informants had to live within a one-hundred-mile radius of Enid, the agricultural center of northwestern Oklahoma. (5) Informants that comprised the financially stressed group had to have self-reported debt-to-asset ratios that placed them in "serious financial difficulty."

Finally, I was a *periodic participant-observer* with one farm family of my life history sample. I followed this family throughout one entire planting and harvesting cycle as a farm hand. My interest here was to investigate changes in the "moral economy of the family" (Collier 1986), or how understandings of the family and family roles were changing as a result of the increasing dominance of industrial values. I examined this issue primarily by looking at the division of labor between farm men and women, the evaluations of women's work and contributions to farm economic viability, and the views women hold about land and how these notions are tied to ideas about kinship.

Since RHP, as an organization, holds a great degree of credibility with the health, social services, and farming communities of northwestern Oklahoma, my access to a diverse group of informants and consequently, my fieldwork experience, was enhanced by this association. But as my fieldwork progressed, I slowly became aware of another source of data—one that I had not anticipated and that had little to do with RHP.

Conducting research with informants who were members of the same society as I am raised some interesting, unanticipated, and challenging epistemological and personal issues. The first one, as highlighted by Kondo, concerned my own positioning—insider/outsider—vis-à-vis my informants. In my introductory meeting with the subjects of this research, I would explain that I was an anthropologist and that I was studying the farm crisis. Inevitably, I would be asked if I was a student at the University of Oklahoma or at Oklahoma State. I would tell them that I was not, that I was a student at Stanford University in California. This, as it turns out, was a very big point against me. For most rural Oklahomans, California is foreign territory. As Stein and Thompson (1992) asserts, Oklahomans' assessment of location is informed by another kind of relative positioning—a psychogeography: Oklahoma is identified as home, as familiar, a place where Christian values reign. The coasts, on the other hand, are characterized as unfamiliar, unknown, as "other," places of questionable values. Upon seeing their faces fall, I would immediately say, "But I was raised in Enid, live here now and work for Rural Health Projects." Their faces would brighten. It is also important to note that I looked, and for the most part dressed, like my informants, who were all Euro-American.

The negotiation process was successful and I was granted insider status to a certain degree. I judged the process a success in the sense that, at the time, I wanted to be accepted and thought that this perceived similarity would produce "better" data. One drawback of their assessment, however, was that they sometimes assumed I had more knowledge than I really did. Eager to feel like one of the group, it was sometimes difficult for me to refute my insider status and ask my informants to elaborate on certain issues raised, particularly concerning the technical aspects of farming.

Slowly, as my research progressed and our mutual trust grew, my informants and I invested emotional energy and expectations into our relationship. To a certain extent, I felt that they wanted their experiences regarding the farm crisis to reach a wide audience. Oklahoma farmers, after all, have suffered a great deal, and the boom in the Oklahoma economy and its subsequent bust produced many blatant injustices that have created suspicion about institutions that have traditionally been considered mainstays of rural life, principally banks and the government. I think that they felt that as one of them, I could be trusted to write an account true to their experiences.

The second issue that arose during the course of my fieldwork was the anxiety I experienced due to the divergence in my own life choices from local cultural expectations. Again, at our introductory meetings, I typically met my informants in their homes, sometimes as far as two hours from Enid. Interestingly, by the time I returned for the second interview, usually a week later, I would discover that they had contacted their network of friends in Enid and had found out all about my family and me.

After an interview one day, Mr. Becker asked, "So your family attends the Catholic Church over in Enid?"

"Yes sir," I replied.

"So what do they think of you being involved in abortion rights political activity?"

Apparently friends of theirs attend the same church as my family and had made them aware of my involvement with the National Abortion and Reproductive Rights Action League, which was frowned upon by many. I told him that my family was very supportive.

I was more than a little surprised by his and other informants' knowledge of my life without my having told them. More importantly, their comments, many times, signaled the ways in which my life choices had varied from local cultural norms. As a result, I felt compelled to begin to sort out more systematically who I was and what I believed—a process

made more difficult not only because I was raised largely in the local culture of my informants, but also because of the guilt I felt due to the disjunction between that cultural ideal and my own experience.

I had done some research in Trieste, Italy, prior to this project. My experiences in the two sites differed considerably. True to anthropological
tradition, it was easier to see everything afresh in Italy and to separate my
own views and biases from the cultural milieu of my study; I liken the process to sorting blocks of primary colors. My experience at home was something else entirely: it was more akin to sorting differently hued blocks of
blue—so similar were my beliefs to those of my research subjects. This
process called into question issues that I thought I had brought to resolution, including, for instance, my own spiritual life and my economic and
political beliefs.

The third issue I encountered was my own stage of adult development
during fieldwork. As Kondo (1986) suggests, it has become quite expected
that anthropologists explain their location or position with respect to their
field site and informants. This discussion usually includes divergences or
similarities involving key issues of power, such as gender, class, ethnicity,
and race. It seems, however, that an individual's life experiences leading up
to research can also affect findings and interpretations.

As I began fieldwork, I was conscious of the fact that I wanted to make
Oklahoma my own, completely. It is the place where I had spent most of
my life and I suppose I was in search of a community—a place I belonged.
Given that I was in my late twenties as I conducted fieldwork and that this
is typically the time when personality integration takes place, my questions
about community were right on time. I thought, given my perceived quasi-
insider positioning, that if I knew more about Oklahoma than anyone else,
that I would clinch indisputable insider status. The process of fieldwork
slowly and painfully made me realize the impossibility of that goal.

A memorable moment of dissolution came as I sat at the Melroses'
kitchen table. Mr. and Mrs. Melrose were key informants and the family
with whom I conducted my participant-observation research. On that
particular day we were discussing historical changes in rural communities,
the consequences of these transformations for social relationships, and
their sense of belonging or community. They told me about the rural areas in which they had grown up, where the exchange of agricultural
labor was required and where social events and interfamily visits were
common—communities in which there simply were more people. They
commented that rural communities are essentially different now. Young

people don't stay because the farm can't support more than one nuclear family. Technology has made it such that people do not need to exchange labor. Mostly, they talked about the competitiveness of their neighbors.

I was dutifully listening and jotting down follow-up questions, intent on continuing this thread of the conversation. My tape recorder was taking it all in. Then all of a sudden, Mrs. Melrose asked me, "Don't you feel a sense of community here, Eric?"

The foot of my crossed leg began to bounce up and down and I began to run my capped pen across my open palm. The recorder documented my response: some gibberish about my migratory family history and phrases like, "but of course my family is here . . . I feel at home here, but I am so different . . ."

The experience left me swimming and for about three months I found it difficult to conduct research, read about my project, or even write field notes. It was not a conscious decision but a pattern of unproductiveness that I noticed after a period of time. An article I read by fellow Cuban-American anthropologist Ruth Behar resonated with my own experience and helped me make sense of what I was experiencing. In her essay "Death and Memory: From Santa Maria del Monte to Miami Beach," Behar (1996) describes her experience of coming to better understand her identity in light of her history which straddled two worlds. Like Behar, my own family history has been an unsettled one: Spain, Cuba, New York, and now Oklahoma. Behar's article brought to consciousness my "project within a project": that fieldwork represented a means to acquire a past for myself. I suppose the reason for the period of unproductiveness had to do with the impossibility of this effort—a failed attempt. While I listened to my informants, I became immersed in their stories of perseverance and courage and realized that they were not my stories or, at least, not my only stories.

The motivation to work again arose from the recognition that my job as an anthropologist was not to document what it is that I find appealing or personally meaningful—that is, my job was not to document lives as I want them to be—but that it required that I confront my ethnocentrism on the most personal level. This is the only way to produce a balanced account, one that allows new and deeper understanding of a culture for its members and nonmembers alike.

Because of some good advice I received from one of my advisors, I learned to use the tension between my own beliefs and values as a relative insider and those of the farm men and women with whom I worked as a source of cultural information. Using myself as an informant, examining the hows and whys of our differences, gave clarity to the beliefs we both held and those that we did not share.

The final issue raised by my field experience was my political position as a "native" anthropologist. While this research held the potential of answering some important academic questions, it was clear that it was my experience of working in rural public health that motivated me to pursue this project.

Once again, my experiences in two fieldwork sites—Trieste and northwestern Oklahoma—differed on this point. In Italy, I had felt like an observer without much personally at stake in the pitched battles being waged by clients, doctors, nurses, and the government concerning the public mental health delivery system of Trieste. In Oklahoma, on the other hand, I felt that much was at stake. I was committed then, as I am now, to local citizen action to improve the social condition. Oklahoma was my home and I wanted to enhance the health of the people with whom I worked. I was *impegnato*—or "committed," to borrow from my Italian friends—to make a difference.

I wish that there were one story of a farm family in danger of losing their land that I could write about that could convey the poignancy of what my work made me feel. The reality is that through my work and research, I was surrounded by stories of individual and social loss. I was struck by how different rural Oklahoma communities were from my naive preconception of rural areas as idyllic settings where the interdependence of individuals shielded them, to some degree, from hardships. What I found instead were individuals in pain, grieving loss—of community, home, land, spouses, and children—and dealing with their own diseases, physical and mental, prompted by the realities of living under sustained and trying social and economic conditions. As a problem-focused applied and medical anthropologist, I wanted to find a solution—I wanted to make it better.

I hope this study can contribute to making it better. I hope this study will challenge the myth that rural areas are idyllic settings in which the interdependence of residents prevents the generation of stress (Wagenfeld 1990). Through this study I hope to provide an in-depth, descriptive analysis of a single setting and contribute to changing the perception of rural America as homogenous (Wagenfeld 1990; Murray and Keller 1991; Human and Wasem 1991). Both of these factors have been highlighted in the mental health literature as inhibiting culturally appropriate interventions. This project examines farm distress and other health issues within their cultural context to illuminate the cultural ideas and structures that underlie health-related behavior. I hope it will help to enlarge our understanding of the cultural barriers that prevent people in rural communities from seeking and obtaining needed services. More importantly perhaps, by focusing on native experiences and interpretations, I go beyond medicine to shed light on structural, political, and economic factors that are affecting our communities in profound and sometimes disturbing ways—to make clear that the most substantial benefits to the public health can be attained only when structural injustices are addressed.

In the chapter that follows, "The Nelsons," I focus on one farm family and describe their story of coming to terms with severe financial debt and the changes this process has wrought for their lives. Through them, I draw attention to the experience of crisis and the patterns of feelings and actions of Oklahoma farm families who found themselves in financial distress.

In the chapter entitled, "Creating Oklahoma," composed of short essays, I highlight critical issues such as the history of northwestern Oklahoma, land as a multilayered symbol, kinship, the evolution of ideas about community, and the globalization of the world market economy. The essays are important in helping us understand how men were positioned within a matrix of cultural meaning to experience crisis as they did and also help us understand the changing cultural context of my informants, challenging some subjectivities while affirming others.

In "The Good Farmer," I begin to breakdown the monolithic concept of the "farmer" to demonstrate the diversity of views and competing discourses on farming, especially industrial farming and family farming. I examine occupational role evaluation to clarify the relationships between farm management, gender, and emotions. In doing so I seek to illustrate the mutual structuring that occurs between these seemingly unrelated domains of human activity and thought and that set the stage—on a metanarrative level—for the individual experiences of crisis.

In the chapter, "The American Agriculture Movement and the Call to Farm," I profile the resistance of this family farm activist organization. Their actions, I will show, criticize hegemonic industrial notions of farming that have implications for a broad array of spheres of social life and which, I argue, subvert and challenge industrial assumptions about gender, emotions, and the hierarchy of status.

In my conclusion, "Modernity, Emotions, and Social Change," I draw attention to the consequences of the growing dominance of industrial values in northwestern Oklahoma and the area's increasing incorporation into the world market economy. I challenge the naturalness of capitalist expansion and explore my informants', and my own, vision of a just society. The concluding section examines the personal consequences of the farm crisis. I call attention to the power of emotions as social commentary and their importance in social transformation.

Finally, I include an appendix in the form of a personal essay called, "Wide, Open Spaces." I include this piece as a response to my informants who asked me about my community. I also include it as a way to share my own breach—of academic discourse, of male proscriptions against emotions—with people who shared so many of their own with me.

They call it desolate
When wind-chased tumbleweed
Lumber, roll, fly
Over flat grasslands,
Wheat fields, paved highways
And dirt roads;
When a clump of half-dead
Cottonwood clings to a
Streambed that fills at flood
But mostly flows
Red clay mud.
I have made home
Driving roads lined dense
With winter straw.
Desolate is not
The place I saw.

—"A Place Called Desolate," Howard F. Stein

Men stood by their fences and looked at the ruined corn, drying fast now, only a little green showing through a film of dust. The men were silent and they did not move often. And the women came out of the houses to stand beside their men—to feel whether this time the men would break. The women studied the men's faces secretly, for the corn could go as long as something else remained.

—John Steinbeck, *The Grapes of Wrath*

TWO

The Nelsons

I REMEMBER DRIVING NORTH. I had finished work for the day. I had just enough time to run home, change, gather my notes and tape recorder, and jump back into the car for the one and a half hour drive. I was hoping to arrive at the Nelsons' home by 7:00 P.M. for what would be our third and final interview session.

It was late spring. I know this because my interview transcript is so marked, but also because I remember the fields; the stalks of wheat were tall, turning from deep green to golden straw. They were reflecting the warm glow of the tired sun and were swaying, it seemed to me, to the press-pull-back of the slow 3/4 time of the classical music on the radio.

The confluence of wheat development and time of year signals my location in the southern plains. Wheat ripens in Texas and Oklahoma first, allowing independent harvest combine crews to plan their work and travel season to begin in this part of the world. From here they follow the harvest, working their way northward and westward: Kansas, Nebraska, South and North Dakota, Montana, and then on into Canada.

Northwestern Oklahoma is austerely beautiful country. The Plains, however, are an acquired taste. It comes with familiarity, knowing how to distinguish a pretty field, with its smoothness, one that allows you to see farther, that invites light and is framed by wildflowers and a few trees to break the wind—from a less attractive expanse that is irregular, obstructs the view with an out-of-place mound, and whose surroundings are not complementary but barren. The plains in northwestern Oklahoma are rolling and often flat. There is little here, generally speaking, that shelters people or animals. I have seen pictures of Enid taken soon after non-native settlement in the 1890s. There were hardly any trees at that time. The ones that currently exist were planted and nurtured, but even these have the look of discomfort, stunted, shy almost, like they know they don't belong.

I am not a believer in environmental determinism, but the people of northwestern Oklahoma do reflect the consequence of landscape and weather to some degree. They are a people who are at the mercy of the environment and—as many would say—of God. Just as temperate rainfall and sun nourish the wheat, severe thunderstorms roll in and sometimes stall, threatening crops and lives with their accompanying hail, high winds, and occasional tornadoes. Perhaps because they are grounded in this potentiality, the people with whom I worked are more direct; there is less art and more substance to social interaction. City people are readily distinguishable when they care to venture our way. At public health meetings, for instance, sometimes comments would be made: "Here come the city folks trying to sell us a bill of goods" or "Why can't they just say what they came here to say?"

People at the mercy of nature also tend to believe; they are a religious people. I would be told, "You can't be in the middle of nature, experience all the cycles of life, and not believe in God. As a farmer you have to have faith that in the spring, the wheat will grow."

Being able to see for miles in any direction is comforting to us. I remember as an undergraduate in the hills and mountains of Vermont, feeling a small anxiety at my center that developed after a time and gradually grew. I was perplexed by its cause and persistence until I realized that I felt cut off from the rest of the world. I was claustrophobic. Space is comforting, perhaps, because it provides perspective; you can place yourself better physically and strategically in what surrounds you. But as I learned from my informants, just because you can see forever in any direction doesn't necessarily mean you know what's coming.

Toward the end of my second interview session with the Nelsons, Mrs. Nelson first broached the subject of their financial indebtedness. "I don't think either one of us saw it coming, did we? We did, I guess, during the tractorcades.[1] We knew that, hey, we weren't getting anywhere. We were not. We were just spinning our wheels and we knew that okay, well, we have this land. That's a commodity that we can mortgage and get more money so that we can achieve more. And so that's what we sought out to do. But that was when the Production Credit Association [PCA] said stop."

Mr. Nelson added, "Because see, we had already built the harbor store system, so we could run extra cattle. We looked at going over and leasing ranch land in Terman County or buying land . . . we needed to go ahead and buy the numbers of cattle to put in and the feed. So we went with Production Credit.

"The tractor that we were using didn't really have the horsepower to do what we needed and so in the course of visiting with them [the bankers], why we said we needed to have a tractor to work with up there. They looked it all over and said—and I think we had right at $100,000 that we were looking at—that it looked good, like we had what we needed to do the update up here, to get it where we could handle the livestock the way they needed to be. We went in, and of course they knew all about how to handle everything. They had all the cash flows and the financial statements and what have you. And we spent weeks putting that all together. They said fine. Your cash flow's working, there's no reason it won't work, go right ahead with it. So boy, they were shoveling us the money!

"We went ahead and built the pens and got the tractor and were ready to buy the concrete bumps to put in the feed lot. And all of a sudden they came running out here and said 'hey, whoa, it's not working.'"

"They wanted you to come in," said Mrs. Nelson. "They wanted to talk to you without me. They didn't want your *wife* to come along."

I asked them why they thought that was.

"Well, they wanted to lower the boom on me," Mr. Nelson answered matter-of-factly.

"I think they were afraid that I was too much of a . . . you know . . ."

"That you couldn't handle it," Mr. Nelson said, finishing her thought.

"Yes . . . yes . . . a woman."

Indignantly Mr. Nelson added, "And here we had worked through this whole thing together and she's always there with me unless there is a reason."

The purpose of the meeting with the bankers was, simply put, to inform the Nelsons that they were on the brink of financial ruin. For the Nelsons, this meeting in the early 1980s set into motion a slow and long-term process of adjustment, of coming to terms with the reality of their financial situation.

I exited the highway that borders the Nelsons' farmland. I made a left turn onto a dirt road and almost immediately took another left onto the short driveway that led to the Nelsons' large, modern ranch style home.

I had first met Mr. Nelson at a focus group session I conducted with another Rural Health Projects researcher. As I have already mentioned, the Oklahoma State University Cooperative Extension Service had recently contracted RHP to assess the problems that farm families in financial crisis were experiencing and to develop support services to address identified needs. During the meeting, Mr. Nelson struck me as being particularly thoughtful—someone who was exceptionally knowledgeable about his community, farming, and agricultural policies and the effects of the farm crisis. His wife, during the course of our interview sessions, proved to be his match as a thoughtful commentator on rural life.

The Nelsons smiled and waved as I drove under their covered entrance. They were warm and inviting and always made me feel comfortable. During our first two interview sessions, I had learned much about the Nelsons' personal history: both are of fairly recent European origin (within three generations). Both sets of relatives arrived with the non-native settlement of Oklahoma via the Land Runs of the 1890s. Both have unbroken farm lineages. They grew up in Oklahoma, were high school sweethearts and attended the state agricultural college; Mr. Nelson graduated with a master's degree in agriculture and Mrs. Nelson with a bachelor's degree in home economics. They have four children who have all attended, or are attending, college. The life they had created for themselves before the crisis is what they had planned on and what they had worked so hard for.

We settled at the dining room table. I began to ask follow-up questions to our previous interviews to flesh out some of their personal history. I did this deliberately. I had learned by now that the subject of the farm crisis arose slowly and tentatively and only after some level of trust had been established. Partly because of the stigma associated with the farm financial crisis, people in rural communities seldom talk to one another about their own experiences; they are not practiced.

Mr. Nelson described what it was like at first, when he first learned of their financial troubles. He said, "I don't think anybody really knew. We didn't advertise it. And with each step we went through, we learned a little bit each time as we went, as to what we were up against and everything. And I think—"

"There was a lot of denial," Mrs. Nelson interrupted.

"Oh, I'm sure. I'm sure," Mr. Nelson agreed.

"Because I can remember you saying, 'What will the neighbors think?' You know."

"Probably was, yeah," agreed Mr. Nelson.

Mrs. Nelson continued, "The people in the community, oh gosh you know. That was early on." Mrs. Nelson turned to me, "Tom had a real hard time early on, from the time we knew we were having problems, 'What was the community going to think?' My answer was always, 'Well, what do I care what they think?' I was more concerned about the family and that our family stay together and be together. You know, the physical aspects of the farm weren't nearly as important to me. But then I think men are so tied to their professions that it's part of their identity."

This last point struck a chord with Mr. Nelson. "Well, and from the time we started farming I had never thought that I would do anything else but farm—and farm until I got ready to retire. And as far as selling the land, I never thought that would be something we would have to deal with."

I asked Mr. Nelson what he thought the community would think if they knew they were in financial trouble.

"Oh, I think the big part of it was my pride. I think it goes back to the peer group. I'm wanting to prove myself, that I can come out here and be a good farmer and do the job and I'm using what the neighbor's doing as a guide—to whether I'm on target or not."

"So it's about the peer group and the fact that you kind of compare yourself to each other? Is that right?"

"Yes. You know, you don't know how much money anybody's got in the bank . . . but everyone looks. If I see a pasture that's been improved, that shows me that someone cares and that they did some research and went out there and tried to improve the pasture."

This brief conversation captures several major themes that I would come to hear over and over again. First and foremost is the issue of pride. The pride of men arose in every single interview session—often. It was appealed to by my informants to explain various actions and inactions, feelings, and consequences of the farm crisis. For me it emerged as the central problematic of this research, the thing that needed to be explained to have a basic understanding of the experience of farm men in severe financial distress. Because of its importance, I reserve a special section on pride at the beginning of the next chapter. There, I use pride as a "hinge" of understanding, which opens the door to the rest of the book.

The second issue raised by this brief encounter is denial. There were at least two kinds of denial. The first kind was highly personal. When families discovered the reality of their financial situation, there seemed to be a period of time before the information "sunk in" and became real. This denial manifested itself as inaction; that is, men would delay taking steps to resolve their financial situation. It seemed that farmers believed that if the problem were ignored, it would go away. In many of the cases that were described to me, the resolution of serious financial indebtedness would have been easier had the farmers concerned acted sooner and accessed existing services that were designed to address their needs. Some never acted. So ashamed of their situation, some never even told their wives. In at least two cases that were described to me, family members found out about their insurmountable debts only when the sheriff came to evict them from their homes.

As alluded to by Mrs. Nelson, women, in general, were less attached to farming; it was less integral to their own identities. Perhaps because of this, they found farm financial trouble less compromising of their own personhoods, less debilitating and, as a result, they found it easier to act. Women were typically the first ones who reached beyond the immediate family context for help. At times, they accessed counseling services or other types of health care. They sometimes contacted the Cooperative Extension Service to learn about programs geared to families in distress. At other times they reached out to extended family members and even to bank officers.

Another informant, Mrs. Perkins, had experienced the entire process of crisis: foreclosure, land loss, and even the death of her husband to a heart attack, which she attributed to the chronic stress caused by the crisis. She said, "The men tend to hold the facts in. They think they can handle it.

There's nothing they think they can't handle. 'I don't want my neighbor to know that I'm failing. I can handle this. I'm not wrong and it's going to be all right.' And the women will pick up the telephone and call and say, hey, we're having problems and here are the facts. And the women will talk to each other. They don't go out and, you know, spread doom and gloom. But they compare constructively, the financial, the economic hardship that they are going through. And it is generally through women that something can be worked out."

Unlike many men who experienced anger at the perceived injustice of their financial situation and at the prospect of losing their land, women told me they were motivated to act by the anger they experienced watching their families suffer.

The second kind of denial involved the community. As the Nelsons mentioned, they certainly "didn't advertise" their financial situation. To be labeled by the community as a farmer in financial trouble meant that one was a "bad farmer" and a "poor manager." Some in the community went even further, attaching a certain morality to unmanageable debt. "I just wonder what they were doing with their time. Maybe doing something they shouldn't have been," mentioned one farmer in the far western part of the state. Others alluded to the Protestant notion that success in life was a manifestation of good and careful living and trust in God. For instance, when something exceptionally good happens to someone, it is common in northwestern Oklahoma for a person to respond, "You must be living right." The logic here is that farm financial trouble is an indication of "wrong" or immoral living.

People were cognizant of the stigma associated with farm financial trouble and, even though individuals responded differently to news of serious debt, this information was held almost universally within the bounds of the immediate family. Troubled farmers in the community behaved normally for the most part, as if nothing had happened. In some cases they continued to make purchases, including large ones (a new pickup, for example), even if it meant they would increase their indebtedness. All this was done to dampen the suspicions of neighbors regarding their financial situation.

The final issue raised by this brief encounter is the role of gender, specifically masculinity, and its relationship to work. Mrs. Nelson, in essence, observes that Mr. Nelson's work is central to his self-definition. Mr. Nelson concurs and adds that his concern regarding the community's knowledge of his economic troubles is centered on his peer group. These comments are in alignment with the assertions of two cultural histories of American manhood. Kimmel (1996) and Rotundo (1993) contend that not only is the increased focus of men on their occupations bound to historical changes concerning work and production—specifically the industrial revolution—but that with this historical shift, masculinity itself becomes a commodity. With the gradual change from "communal man" to "self-made man" during the latter half of the nineteenth century, men entered, and began to link their identity to, the volatile marketplace and notions of economic success. Aggression, competition, and self-interest became perfectly acceptable traits associated with middle-class men in their quest for economic success and power. Manhood was no longer a given, but could be won or lost. Mr. Nelson's concern with his peer group is echoed by Kimmel (1996:26–27), who asserts that, in large part, it is other men who are important in gauging one's masculinity and success. Both concepts are largely defined by men's relations to one another. Masculinity is in part, then, a same-sex social process.[2]

It became clear to me that what was at stake was the individual's status within the community and his reputation as a farmer, breadwinner, and provider for his family. What emerged, too, was a picture of a society that was highly competitive. Success as a farmer was measured in terms of the amount of land owned, the condition of the land (maintenance, repair and general upkeep and appearance of a piece of property), and yield of wheat per acre. Most of this information was readily available to all since it was literally out in the open. Revelation of a farmer's financial trouble compromised his position in the hierarchy of status.

Perhaps farmers withheld information regarding their financial status from those beyond the boundaries of the immediate family for additional reasons. The strong stigma attached to farm financial trouble harkens back to Dudley's (2000; 2003) notion of the entrepreneurial self and the morality that it supports, one that radically individualizes even as it requires individuals to conform to community norms. Financial failure betrays this morality and those who experience it incur social censure if their situation is made known. These farmers may feel "cast out." Mr. Nelson said, "Because even in our family . . . cousins and what have you . . . and they were [sitting] in a group. If I went over and sat down with them, they'd all get up and move. And if I wouldn't go over there, they'd just keep talking and keep everything going. But you know there were problems that way.

"My godfather was there. He is one that . . . even though he knew the situation that we were in, he would always complain, 'Well why should I have to pay a higher interest rate because the guy over there has filed bank-ruptcy and he's trying to hold onto his? Why don't they just take it all away from him and just put him in jail or whatever it takes?'"

"Well, and Tom is also on the ASCS [Agricultural Stability and Conser-vation Service] County Committee. He was chairman of that group," Mrs. Nelson added, turning to her husband, "and you had never, ever had a time when you weren't elected chairman until after the whole thing came about." Addressing me once again, "And then all of a sudden he wasn't elected chairman anymore. And even . . . it can be a little subtle, but it is still there. And you hear that, 'Oh, bankruptcy isn't such a big deal any-more.' Why, it's not a disgrace like it used to be. It's still there. Still."

"I finally resigned off of that board," Mr. Nelson said. "I decided that I couldn't make decisions for myself, [so] how could I make decisions on a county-wide basis for other farmers?"

As this statement indicates, farmers participated in their own social iso-lation. Men in crisis would resign from publicly elected office, school boards, the local cooperative board, and other committees. Similar reason-ing was always used: "If I can't make decisions for myself and my family, how can I make decisions for other farmers and their families?" This logic reveals the degree to which farmers internalized failure and dismissed macroeconomic and policy factors in assessing their own occupational role and effectiveness.

Farmers even withdrew from churches. I had imagined, before I began to research the topic, that religious organizations would be seen as a source of support and help, but this was not the case. One minister said to me, "I know when a family is in trouble. The first thing that happens is the man will stop coming to church." I asked my informants about this phe-nomenon. They agreed and asserted that churches, in essence, are social institutions like any other. They are simply an extension of the social fab-ric. In other words, church involvement represented social censure, not an escape from it.

Part of the logic behind the social isolation and censure farmers experi-enced in times of crisis may be revealed by this story that was told to me: It was common knowledge in the community that Mr. Bennett was in severe financial trouble. He was well known, being the church organist. One day, one of my informants, Mr. Ross, who had been trained in crisis interven-tion, received a call from Mr. Bennett's wife. She said her husband was in the basement; he had a gun in his hands and refused to come upstairs. Mr. Ross raced to the house to meet with him. They spoke at length. During the course of their conversation, Mr. Bennett "broke down" and began to cry. He said that this was the first time anyone had acknowledged his situation. Mr. Bennett asked Mr. Ross why no one had ever talked to him before, why no one had ever broached the subject of the troubles he and his family were experiencing.

When I asked my informants the same question, my query was greeted with almost primal sounds and gestures: grunts, humming noises, and shakes of the head and shoulders. They finally were able to articulate that they would not know what to say to a man in such a situation. If they did engage him in conversation, they told me, they would be embarrassed to see him become emotional. Thus, one potential reason for the isolation and censure many men experienced may be due to an avoidance of the discomfort caused by witnessing emotions deemed inappropriate for a man.

The isolation of men in times of financial trouble went even further: men withdrew not only from the community, but also from their own families. Stories told to me about this process caused me to imagine a scene that has haunted me ever since: that of a man sitting alone in the dark in the living room of his home looking out of a window at night when everyone else has gone to bed. This withdrawal, men explained, was due to the profound sense of failure they felt on all levels: individual, familial, and societal. They experienced self-blame and were self-critical about not being able to fulfill one of their perceived principal responsibilities—that of being financial providers for their families.

As a result many men simply kept to themselves; others abused chemical substances. In addition, domestic violence, child abuse, and suicide among farm families in crisis have been reported in the literature (Jurich and Russell 1987). Other men worked themselves to death—sometimes literally—to better their economic circumstances. What is clear is that men's withdrawal and their destructive coping strategies significantly affected family dynamics.

As the issue of familial relationships emerged during the course of my interview session with the Nelsons, the conversation became passionate. It was evident that my questioning prompted ideas and feelings that had not been previously shared by Mr. and Mrs. Nelson. Not trained as a therapist, I was concerned that I was in over my head. I became worried that I would not be able to respond thoughtfully to their remarks. Also as a researcher, I wanted to ensure that our conversation was not hurtful.

"He was like . . . like I said there was a time that he wouldn't talk to me at all and you know I was, well . . . I just turned away from him," said Mrs. Nelson.

I asked her how it made her feel.

"Well . . . sad. It made me feel insecure about myself. I felt like hey, if we've got our family and we have our health, why can't that be enough." She turned to Mr. Nelson, "Why can't that be enough for you? And it wasn't enough." She turned back to me and continued, "I guess throughout all of this, the thing that nagged at me more than anything was that Tom loved the farm above everything: above me, above the family. The farm came first."

During this conversation, Mr. Nelson was staring at his hands folded in his lap. It seemed to me that he was very respectful of his wife's perspective. He said, "I didn't realize that I was doing it this way. I was out here farming as hard as I could. There was no doubt that I was neglecting our marriage."

Mrs. Nelson continued, again addressing her husband, "Well it's just that your need for the farm was so strong, that I couldn't supply what it would give you and the loss of it was such a devastating thing for you to cope with, that I was at a loss to know what to do. I just was. And well hey, our marriage went into a tailspin. It did . . . we were a hair's breath . . ."

From divorce. The Nelsons did not divorce. They did seek counseling, though they did not feel it was particularly helpful, and have remained committed to working out barriers in their relationship that they attribute to the crisis. According to therapists I interviewed, the patterns of behavior that I have described often led to divorce: self-blame on the part of men, Herculean work efforts to save the farm, men's increasing isolation, lack of communication, and consequent feelings of loss and grief on the part of women.

But there is a dynamic here that is hinted to by Mrs. Nelson's comments, one that is not fully captured by the language of counseling. It is this: if a woman perceives that her principal responsibility is to be the helpmate of her husband, her inability to effectively support him corresponds to his inability to save the farm and provide for his family. Both are moral failings of the deepest kinds: hers as a wife and homemaker, his as a husband and farmer.[3]

One of my informants, Clara McCaffrey, 49, had divorced her husband after years of fighting off foreclosure and bankruptcy. She said, "You know, there's a great sense of failure for someone like me who believed, you know, in better and worse, richer and poorer, and sickness and health. You know, that's what I thought it was you did and I thought—I guess, I thought a little bit of not going through with the divorce for a while but, you know, some other things were going on that made me decide to just go ahead and go through with it. I can't really say that I experienced a lot of depression that a lot of people talk about going through with a situation like that. As time has gone on, and I've reflected on that, I think there may have been a couple of reasons for that. One of them was, I think, a lot of my depression was over. I had been depressed before. I had been smothered and suffocated and confused and lonely and all those kinds of things in the marriage that I just saw dying—in the way of life that I saw dying. So when I finally had, when I finally broke loose from it, it seemed like it was kind of a breath of fresh air."

Mental health professionals that I interviewed agreed that this was a fairly typical pattern: women would have begun to mourn the loss of the relationship some time before it actually ended and, by the time a woman would announce her intentions to her husband, she would have completed a substantial amount of the grieving process. Men often felt broadsided by the news of an impending divorce. They had been so consumed with keeping the operation afloat that they had no idea that the collapse of their marriage was in the offing. Men, too, felt that their hard work and inattention to the relationship were justified since they were struggling to maintain the financial solvency of the operation for the sake of "the family."

Naturally, the effect of the crisis was not limited to the married couple; children were profoundly affected by stress as well. As part of our assessment process for RHP, we conducted a focus group session with teenagers in the Oklahoma panhandle. We chose a high school that drew students from a very large, sparsely populated rural area. The principal and school counselor helped us to recruit young people who were representative of the school's various peer groups.

The evening of our focus group session, the invited teens entered the high school cafeteria raising a ruckus with their laughter and banter. They were a very friendly group of young people, receptive to us and to our questions. I was struck, however, by the change in mood of participants— from jovial to somber—as we proceeded through our open-ended questions. True to form, the subject of the crisis and its effect on students and their families emerged slowly.

The high school students spoke to us about the unwritten family rule prohibiting discussion of the family's financial troubles outside the home. They described their feelings of loneliness and depression caused by their perception that their family was unique, the only one in the "whole world" that was experiencing this hardship. They also told us that their parents' stress level affected them. The teens said that they sometimes felt like a burden to their parents. Some responded to this feeling by diligently working on the farm, doing their chores in a timely fashion, or by simply saving money—foregoing a new outfit or the latest pair of brand-name sneakers.

Interviews we conducted with teachers revealed that other students rebelled under the pressure by drinking or being disruptive in class. Still others were overwhelmed by the strain and experienced almost catatonic depression. Teachers reported to us that students' grades, almost universally, would begin to plummet.

Young people did want to talk, they told us. They wanted to talk to one another about their experiences. One young woman described her intense feeling of relief and almost joy when she finally decided to tell her best friend about her family's financial trouble. She discovered that her friend's family was in the same predicament. Talking, she said, helped her to feel less alone.

Teens also wanted to talk to their parents. They did not want to be kept in the dark about the family's financial situation. The lack of communication with their parents caused them to speculate about their futures, assume the worst and, as a consequence, become even more frightened: Were they going to be kicked off the farm? Where would they live? How would their parents make a living?

I remember the other researcher and I left the focus group feeling that these young people needed help coping with their situation. We spoke to their school counselor to assess the likelihood of an ongoing support group and contacted the community mental health center to alert them of our findings.

But the predicament of young people has not gone unnoticed and help does exist in several areas of the country. One excellent example is a curriculum developed by Sue Schlichtermeir-Nutzman of the Community Mental Health Center of Lancaster County, Nebraska. It is called *Seeds of Change: Growing Up On Today's Family Farm*. The lessons are intended to be integrated into the regular school curriculum and to engage students in activities that promote parent/child communication, encourage young people to talk about their feelings, and teach strategies for staying healthy and coping with grief and peer pressure. The last page of the curriculum vividly conveys the cultural proscription against sharing information about the family's financial and social situation with others. It contains an illustration of a young girl (who had been naughty, one assumes) standing at a chalkboard writing over and over again her punishment—what she must learn: "I will not talk."

The Nelsons, unlike many families, felt it best to talk to their children. Mrs. Nelson said, "we tried to keep them informed, but then . . ."

Mr. Nelson joined in, "That was a decision . . . when we realized the problems that we were in with the farm, we did, we sat down with them and we started talking to them about it. And probably, we were saying, we don't know if we'll be on the farm [in] a year or so. And of course that upset their applecart then. But like I say, we didn't realize it then, but we do now, looking back, that they were so concerned too that they were having trouble holding their lives together . . . the insecurities of not knowing what was going to happen."

Mrs. Nelson elaborated on this point, "We had two kids in college at the time and our youngest son was in high school. We got in the car and went down to Kelliton to the college and took the kids out to the park and had a little discussion with them about the fact that we had filed for bankruptcy. Then those next two semesters were really tough on those college-aged kids. They had problems in their classes, problems . . . Well, finally our son was having so much stress with his classes . . . his second semester why, he enrolled there maybe a week, and was breaking out in hives. And so he said, 'Hey, mom I can't continue.' I said 'OK Sammy, you're the one that's important, you're the only one who knows how much you can stand . . . why don't you just withdraw for a while . . . whatever you want is OK with us.'

"He just got his B.S. degree. He is going to be twenty-six in June. He got his B.S. last May. I don't think he knows what he wants to do. However, he is involved in farming with Tom and by himself now. And I don't know how long it's going to last whether he's going to stay . . . He was involved several years ago with borrowing money to buy cattle, put on wheat pasture, that particular thing . . . just . . . just . . ."

"Worried him to death. The whole time," Mr. Nelson added.

With time, many families would learn to manage under the pressure and begin to more effectively assess their options. Periodically, however, farm families would be confronted with difficult decisions, obstacles, and hardships that would challenge their ability to cope.

One of the principal difficulties that farm men, in particular, faced was the loss of personal control and power as they tried to manage their overwhelming financial situation. Their role as decision makers for their operations was what, in many ways, defined them as farmers—defined their own sense of identity and their personhood in the consciousness of the community. Mr. Nelson spoke of his lack of preparation to confront his

new circumstances and the jarring nature of his forced personal redefini-
tion, "But that's how this whole thing has been. It's been a process that we
didn't look forward getting into. And we were never schooled about anything of what it would be like. So we're depending now on the lawyer to
work us through with the bank and the court and everything. Almost like
being led as a child. I almost feel myself down here holding on to his hand
to go through it. Because there's no way that we would have known or
been able to go through any of this anyway."

But once bankers had identified the problem of unmanageable debt,
farmers were essentially at their mercy and that of the courts. Farmers' options were limited, further compromising their sense of self-determination.
One obvious alternative would be to leave farming altogether: sell the land,
the machinery, and the house, move to town and get a job. Unfortunately,
this was easier said than done. "It is expensive to get out of farming," I was
often told. The costs associated with farm operation were so high, that depending on current real estate and farm equipment prices, there was no
guarantee that one would be able to extinguish one's debt even if all the
property and machinery were to sell at reasonable prices.

Also, leaving the farming profession was simply not an option for many
men. Farmers were at their most impassioned when discussing their "calling" to farm, their life's work, the thing they felt they were meant to do.
Mr. Nelson commented, "My interest through this whole thing is the fact
that I started out farming and I've wanted to farm, and I've wanted to fight
it to be able to stay on the farm. And that's still my whole goal, is to hold
onto it. I just can't seem to turn loose."

Mrs. Nelson added, "Well, and that too was a real source of tension because Tom will hang onto this land 'til he has one finger left out of the
grave. He will. He will not give up. He's not going to throw in the towel.
He will do whatever." I had been told of farmers who, upon realizing that
their finances were deteriorating, would sit down and calculate the extent
of their loss per year in order to determine how much longer they would
be able to continue farming. A certain "rationality" is no doubt involved in
this calculation, but not one that would likely be well understood by most
agricultural economists.

Another obvious choice for families in crisis was to seek additional income from off-farm employment. This option also contained inherent difficulties. Partly due to the continuing declination of the rural economy, the
communities in my study were witness to significant out-migration—a demographic reality that held little allure for new industries in search of a
sizable and skilled workforce. Consequently, for many of my informants,
nonagricultural employment opportunities were limited.

There were other impediments as well. For some, available jobs required that they travel great distances. For others, the low pay of obtainable positions did not justify the extra time spent away from the operation or home; my informants felt they could achieve the same financial result by becoming more vigilant of cost-cutting opportunities at home and on the farm. Mr. Nelson added, "As far as getting a job, if a woman is going to work and she is getting minimum wage, by the time she pays for her oil and gas to get to her job, she's not going to bring very much home. So she's really almost got to have a pretty good high-paying job to make it worth the trouble of even going. And if she's got young kids at home it's worse yet, 'cause she's going to have to pay a babysitter sometime, somewhere." The men and women that I interviewed, the majority of whom were college educated, also felt that, after having farmed for so long, they did not have the marketable skills required in today's workforce since it had been so long since they had "used their degrees."

Banks or courts often dictated changes in farm operations or made demands that consumed families' coping resources. These were often very difficult for farmers to contend with because they entailed some form of public disclosure of the family's financial situation. For instance, a farmer could sell land with the hope of improving his financial status. However, there are few reasons a working farmer would sell land—the principal one being that the farmer is in financial trouble, I was told. Consequently, putting farmland on the market is tantamount to a public acknowledgment of this fact. One man related the story of a fellow farmer, who after much soul-searching and deliberation decided to sell some land to ameliorate his worsening fiscal condition. The farmer took out an advertisement in the regional newspaper. He sold his land and his financial state improved. More significantly, this individual had several other farmers confide in him that they wished they had the courage to do what he had done, namely sell their land.

Driven by the bank, the Nelsons did pursue this option to decrease their debt burden. They sold land that Mrs. Nelson had inherited as well as property they had acquired since their marriage. The Nelsons agreed that these were among the most difficult times they faced. The bank had imposed deadlines for the sale of the land. The pressure was on and Mr. Nelson was feeling it.

"Those were real tough, emotional times, through all of that. Because, to look back on it now, I was in no shape to make any kind of decisions. None whatsoever," he stated.

Mrs. Nelson explained, "Well Tom was bound and determined that the bank had an appraisal of the property and since several of the parcels were to go to his brother-in-law . . . he was bound and determined that his brother-in-law wasn't going to pay more than the appraised value. Is that not right?"

Mr. Nelson agreed, "Well, I think the whole thing was that they had the appraised value, but I could not . . . understand that even though that was the appraised value, if we sold the property for more than that, the money would go to the bank and pay off part of the loan that we still have now."

"Yes. And the attorney and I both tried to reason with him. Well hey, that land could have sold at . . ."

Mr. Nelson finished her thought, "We probably could have gotten $40,000 more . . . out of the properties that we sold—at least that much. That would have reduced our debt that we are still looking at hanging over our heads now. But like I say, at that time . . ."

"He would not listen to reason."

"No. They could not get through to me at all."

"So you sold it at the appraised value?" I asked.

Mr. Nelson answered, "Well, that's what we ended up with. Yeah."

As this conversation indicates, Mr. Nelson's capacity to make sound economic decisions was impaired—further challenging his sense of self and identity as a farmer.

The sale of inherited land was more emotionally arduous than the sale of purchased land—a point to which I will return in the next chapter. Implicated in this phenomenon is a strong association between land and notions of kinship and memory. Because of this bond, hard feelings arose between the Nelsons when it became obvious that the sale of Mrs. Nelson's inherited land would be necessary. Partially in response to the experience they had just had, Mrs. Nelson decided to take control of this venture.

Mr. Nelson commented on the wishes of Mrs. Nelson's parents on the matter, "We were pretty much told that it wasn't to be sold."

Mrs. Nelson elaborated, "Yes, and that was the hard part, telling him [her father] that we would have to sell it. And of course they did everything in their power to see that we didn't in his lifetime. But after he died, in the natural course of things, we ended up having to sell it anyway. But we really haven't had to sell your family land."

Mr. Nelson responded, "But that bothered me too. Because what we did . . . my family land was in this area. Of course this is where we live. And hers was over there by Fielding, which is, what, about fifteen miles from us? So we talked and talked and talked but we were going to have to sell some ground and since that was the furtherest away, that would be the one to go because we had to run clear over there to farm it."

"I guess I had antagonistic feelings in that regard because why should I sell my land when this over here . . . you know. It was our problem. Why should my family land be brought into it?" Mrs. Nelson reasoned.

Unfortunately, circumstances—a hurriedly organized and ill-timed sealed bid process—once again interceded to result in a less than optimal financial return.

Banks did not stop with the sale of farmland. They often forced farmers to sell agricultural machinery and other equipment at sales that would be held on the family property. To encourage attendance, these sales were announced in the local newspaper. To many farmers, this was the ultimate in humiliation. In essence they felt that, literally, an advertisement had been placed to announce their failures as farmers.

It was clear that farm sales were important events in my informants' minds. Family members traveled great distances to be together for the day. Two of my informants videotaped their sales. They were universally dreaded events.

The sales themselves, however, at times did not meet my informants' expectations. A few farmers said that, as much as a sale, it was also an opportunity for the expression of community support, something they had not anticipated. It is my conjecture that some farmers experienced relief because the "family secret" of farm failure was finally out in the open and that its reception by the community was not nearly as negative as they had imagined.

Mrs. Nelson commented, "We didn't have it right here on this place, but we had it across the highway. I thought that it would be a horribly strenuous time and it turned out not to be nearly as bad as I had anticipated."

I asked them why that was.

They had decided, Mrs. Nelson responded, that "on the day of the sale we were going to all be over there as a family and not me stay at home 'cause I'm the 'little woman' type thing and Tom be the only one over there. And we had all of our kids there. But the people who came were just as nice as they could be. They couldn't have been kinder. I felt nothing but friendship."

Mr. Nelson added, "Well and so many people that we have talked to . . . said how people came in as vultures and . . . be glad that you'd gone under. We didn't find that at all. In fact, we were proud of the people that did come because we were trying to get as much for the machinery and equipment as we could."

I asked, "Why did you think it was going to be a bad experience? I mean, what made you not want to go there, either of you?"

Mrs. Nelson responded, "Well, it's just really hard, correct me if I . . . it's just really hard. We had accumulated the machinery all through the years and all of the pieces meant something to us. It's like Tom is always saying, he'd buy a certain piece of machinery for Mother's Day or my birthday or whatever through the years you know. It was part of the business—it wasn't just Tom's business, but I've always felt a tie to it too . . . it was part of our lives."

Farming is still a part of their lives. The Nelsons did not lose all of their land nor were they evicted. Through a Chapter 12 bankruptcy settlement agreement, which entitles individuals to continue to make reduced payments on part or all of their debt, they were able to stay on the land, and, for the time being, continue to farm.

"Well now . . . we live more month to month, because we are on a very tight budget . . . at least that's what they tell us . . . with the bankruptcy judge . . . from what we have decided recently is that as long as we make two payments, one in August and one in December, they really don't care what happens in between. So we're always, 'Is this right? or is *this* right?' or what should we be doing? And still we don't want to do anything that would jeopardize the plan. I mean we've got an agreement with 'em. But we're still kind of in a fog all the time."

It is important to underscore that severe financial debt is usually not addressed overnight, but is a process that can last for years. The chronic nature of this stress has had, and is having, grave consequences for Oklahoma farm families.

The most disquieting and publicized effect of the farm financial crisis on families has been the issue of suicide. A study conducted by the Oklahoma State University Cooperative Extension Service found that suicide was the leading cause of farm-related death among farm men. Farmers were five times more likely to die by suicide than by agricultural accidents, previously thought to be the leading cause of agricultural fatalities.

In response to the mental health needs of farm families, the Oklahoma Conference of Churches established AG-LINK, a crisis hotline, to field calls from rural areas. The program has a licensed counselor to answer calls and has developed a peer-helper system throughout the state that dispatches volunteers to homes of families in crisis when time is of the essence. Mrs. Perkins, the director of the hotline program while I was conducting fieldwork, estimated that therapists made more than 150 on-site interventions with farmers every year within the state, a state that had the third-highest number of farm suicides, behind only Montana and Wisconsin (Dyer 1996). Northwestern Oklahoma, Mrs. Perkins told me, is the hardest hit area of the state. This is due to the proportion of full-time family farmers in the region, the highest in the state.

Even so, Mrs. Perkins believed that the suicide numbers were dramatically underreported. She stated that there are the official numbers derived from the Oklahoma State Health Department, Division of Vital Statistics, and then there are the unofficial numbers: "Many of these suicides are turned in as heart failure or some other reason. Natural causes, whatever. Because of the doctors being in the local community and being close to the families, to help the families. I can understand that."

During my time in the field, it was clear that farm deaths sparked considerable speculation among my informants—two in particular. One man was working out in his field performing some minor repairs. He apparently passed out under his truck, the story went, and was asphyxiated by the car exhaust. Another man, reported his family, had been killed practicing his roping while riding his horse. Somehow, they explained, the rope had been caught around his neck and his horse dragged him to his death. Some of my informants found these stories unconvincing.

"We've had several that called it an accident. Fifty-four year[s] old and worked around horses all of his life and wound up that a horse dragged him to death. Had a half-inch around his neck, tangled up. I know it can happen . . . but not to a fifty-four year old man that has handled horses all of his life! He was heavily leveraged and fought it for three or four years . . . it ate him up alive, but you can't truthfully say he killed himself. But I am satisfied it contributed to it," said Mr. Gains, a family farm activist.

Suicide deaths may also be underreported because they could be masked as agricultural accidents. Farming is considered one of the nation's most dangerous occupations and agricultural accidents are fairly common. It was not terribly unusual for me to have informants who had experienced serious mishaps on the job. Agricultural accidents provided cover for farmers, reeling under severe strain and depression, to commit suicide and still enable families to collect insurance benefits to pay off debts and stave off land foreclosures. Many farmers were conscious of this option; it was chilling to hear more than one tell me, "I am worth more to my family dead than alive."

The reality is that during periods of economic decline in the agricultural sector, farm accidents do indeed increase in frequency and their incidence mirrors that of suicide. Farmers' lack of funds to properly maintain and repair farm equipment may provide another explanation for the rise in agricultural accidents in periods of fiscal uncertainty; thus the deficiency of resources may serve as a confounder, obscuring the true nature of the relationship between suicide and accidents. A psychologist, heavily involved in the care of rural families, gave me still another explanation for the increased incidence of farm accidents. During times of emotional depression and stress, he said, farmers may not intentionally be self-destructive, "but they pay less attention to living."

Even though many experienced extreme anguish, farmers reported they were reluctant to seek assistance. One factor inhibiting access to health and social services was the stigma farmers associated with mental illness, a phenomenon that was not only widespread, but also deeply resonant. This sentiment may be due in part to a lack of information; many of the men and women with whom I worked admitted that they simply did not understand mental illness. They most often associated it with the most debilitating conditions (e.g., unmanaged schizophrenia, bipolar disorder, and mental retardation). In fact, the mental health system may be unwittingly perpetuating this stigma by concentrating its services and resources so narrowly on the severely and persistently mentally ill. As a result there is little public awareness about other mental health initiatives, such as those programs designed for people experiencing situational stressors, depression, or anxiety.

Another obstacle to access in northwestern Oklahoma may be the high value placed on the personal qualities of rugged individualism and independence (qualities which, as we shall see in the next chapter, have historically served the region's inhabitants well). Many farmers understood the need for services as an overt expression of debilitation, weakness, and dependency. Perhaps this is why the admission of hardship requiring mental health care was talked about in such dramatic terms by my informants. One man equated it with finding out that one had acquired HIV, another with bankruptcy.

Seeking mental health care also represented the potential loss of freedom; my informants said that they were afraid they might be locked up if they sought therapy (ironically, if they were to seek treatment and would happen to mention that they had thought about suicide, the possibility of forced detention would be very real). It is understandable why they would be reluctant to seek services since the state hospitals, they perceived, were little better than the "snake pits" of the 1950s.

Visiting one's family or general medicine physician is a slightly more acceptable form of help-seeking. But again the choice of when and whether to seek medical care is largely culturally conditioned. Stein (1987b), drawing upon his experience as the social science curriculum coordinator for the University of Oklahoma Enid Family Medicine Residency Program, illustrates that not only is health seeking tied to the agricultural season (pre-harvest and harvest being the periods of lowest health service utilization), but also that men will only go for help when their bodies "break down." Farmers' definitions of health and illness are not tied to phenomenological experience, but rather, "machine-like," to bodily performance, particularly in relation to work (Stein 1992:8). In other words, they consider it appropriate to seek health services only when they can no longer perform essential job functions.

If medical care is sought during periods of financial and emotional upheaval, Stein asserts, "the physician might in fact be the only person whom the patient, spouse, or other family member, feels 'safe' enough to entrust sharing these devastating feelings—feelings that are often couched in the form of physical symptoms or complaints" (1992:8). In addition, Stein continues, they may minimize their symptoms and assert that they scheduled an appointment only to appease another family member's desire (1992:8).

Other obstacles to care also need to be considered. Because of their serious financial situation, many families cannot, or choose not to purchase health insurance. As a consequence, a number of farmers simply cannot afford care or avoid treatment because they are terrified at the prospect of generating yet another bill. Under these circumstances, routine preventive health care becomes illusory.

The family physician may pose another barrier. Medical school training in psychological and sociological aspects of illness is of little consequence. Medical students do not receive sufficient training to effectively assess the psychological status of patients and may not even be capable of providing appropriate referrals. Also physicians, as members of the same society, may hold similar biases concerning mental health services and may not be comfortable discussing psychological issues with their clients. Additionally, they may view mental health care as not sufficiently "scientific" and thus less valuable.

Another barrier was my informants' characterization of many health and social service professionals as outsiders. They believed that physicians and therapists, particularly those not raised in the communities in which they practiced, could not possibly understand the plight of farm families: the complexities of running an operation, the importance of the farm in the life of the family, and the dynamics of rural communities. This conceit is partially based on farmers' experiences, but also, I would argue, on the mystique of farming and the farmer as "unknowables," a notion grounded in an agrarian discourse employed by many family farmers to distinguish themselves and their practices from those promoted by the "science" of industrial farming, a topic I will discuss later.

Cultural conceptions of masculinity and male roles are at the core of many of these barriers to assistance. Men cannot demonstrate need of any kind, according to my informants. To do so would convey weakness; public knowledge that a farmer may not be in control would compromise his position in the hierarchy of status. Again, health professionals, as members of the same culture, may also hold to these notions about gender and, as a result, may be embarrassed to reach out to men and families who require assistance. Because men in crisis often withdraw, due to their perceived failure to live up to the cultural ideals of manhood, there may be little social support available due to the cultural presumption of male self-sufficiency.

I have told the Nelsons' story to convey a sense of the cultural factors that can condition the experience of crisis—including economic decision making—and leave many families in isolation. If one word could express what men and their families felt as a result of the crisis, it would be betrayal. The crisis caused a profound shift in the Nelsons' thinking concerning institutions and values they held to be fundamental to their lives. Their belief in government as a protector of the people, in hard work as the principal means of economic advancement, in the unwritten social contract, and in banks as ethical and fair institutions were all compromised as a result.

Mr. and Mrs. Nelson regard the future with uncertainty. Toward the end of our interview, Mrs. Nelson said, "I'm 54 now and in about ten [or] eleven years, and if they raise the Social Security age that means a couple of more years, and we may not have the land—it's certainly not going to be paid for by that time, unless the economy turns around and we have some income. From what the legislature and representative people say, there's only two percent of us out here farming, but what does anybody in town care about what's going on out here? Now when they don't have food on the table, then they'll get concerned. I was reading that a big company just closed and they laid off 2,400 people. Since back in such and such a year we've taken 500,000 farmers off the land and no one's even paid attention and could care less about what's going on."

In the following chapter I begin to make cultural sense of the actions and inactions of farm men in light of this profound social shift. I use pride—culturally informed and individually experienced—as an analytic tool to understand the relationship between emotions, gender, and social action.

Social systems must operate to structure the
psyches of both sexes to produce their desired
ideal types, and they do so by controlling what can
be thought and felt.

—Jill Ker Conway, *True North*

Fields that mean more than fields, more than life
and more than death too.

—Edna O'Brien, *Wild Decembers*

It is 1971, and Mirek says that the struggle of man
against power is the struggle of memory against
forgetting.

—Milan Kundera, *The Book of Laughter and Forgetting*

THREE

Creating Oklahoma

Positioning Farm Men for Crisis

THE BECKERS, both in their late fifties, were unusual in my sample of informants in that they did not own their own land. Rather, Mr. Becker was the manager of the extensive land holdings of a wealthy family that lived in the city. If anything, this structural position lent their perspective a degree of disinterestedness.

I always enjoyed visiting with them in their home. One evening, after dinner, we had the following conversation regarding the emotion of pride and its shadow, shame. We were discussing farmers and the diverse strategies they use to conceal their financial troubles from the community at large. I asked Mr. and Mrs. Becker why they thought that farmers felt compelled to hide their financial reality.

"Pride," Mrs. Becker said.

Mr. Becker added, "Probably the same reason you keep any problem . . . if your children have problems or if you have . . . it is a matter of pride. You try to keep those things to yourself. In a lot of cases you don't want . . ."

Mrs. Becker jumped in, "It might be pride, but it could be too that this is a private issue, you know and don't want everybody . . ."

Mr. Becker continued, "There's also a matter of shame."

I asked, "Could you elaborate on that for me?"

Mr. Becker answered, haltingly at first, "Well, its hard, really hard to . . . there are people who . . . let's say I'm . . . I had all these ambitions and we're talking about a large group of people. Farmers became very ambitious during the good times and, by golly, [we're going to] rake in all this food and feed the world and they're going to pay us for it. So we're going to buy lots of land. We were very ambitious and we were . . . then everything went bad, prices dropped, interest got outrageously high. People had a hard time admitting to themselves, also to the bankers, that they were in trouble and that they had failed. We could blame ourselves, we could blame the government, we could blame our bankers."

Mrs. Becker added, "In a lot of cases the being ashamed part would come from if you got land from your parents, inherited land, and then ended up losing it. That's something to be ashamed of."

I asked her, "Now why is that so shameful?"

"Well, they worked hard to get it," she answered.

"OK, assume I don't know anything about this at all," I said, encouraging her to continue.

She responded, "Well, now this didn't happen to me, but if my parents worked real hard to . . . you know they pioneered, they struggled so hard to get this land and they handed it over to me and I can't hang on to it. I can't hang on to the whole place or something. The idea I think is that I can't do something. And it just made people ashamed that they couldn't . . . you know it's one thing to go and buy land and not be able to pay for it, than to have it given it to you and then have to mortgage it and then lose it."

Captured in this brief exchange are several of the factors that helped position men to experience the farm crisis as they did. For instance, Mr. Becker mentions the "good times," or the boom period, in Oklahoma's agricultural sector that spurred many farmers to expand their operations, making them especially vulnerable when agriculture "busted" in the early 1980s. Mrs. Becker discusses the shame of losing inherited land—and making a clear distinction with bought land—suggesting an association between land and kinship. Finally, the emotion of pride is premised as the reason men would want to hide their financial situation when in crisis; the emotion of shame is cited as a culturally understood response to losing inherited land. Both signal the complexity of the evaluations farmers make regarding their position in the web of social relationships that constitute their communities and families, present and past.

As I mentioned in the last chapter, in every interview session I conducted, pride was mentioned often. The emotion of pride was used by both male and female informants but always in relation to men. Pride was employed by my informants to explain social action that was generative and positive—as a force of production: a man's diligence in his work, his careful nurturance of the land, and maintenance of the family property. It was also utilized to explain his role in the community and his leadership positions in local organizations, church, or in elected office.

Pride, however, was more commonly used to account for action or inaction that could be understood as destructive and negative on the part of farm men. For example, it was used by my informants to justify a man's refusal to seek assistance once the discovery of financial trouble had been made or to explain a man's withdrawal from the community if a farm sale on his property or his bankruptcy notice were made public. It was also implicated in the low-level utilization of health and mental health services by farm men. Finally, as we have seen, pride was offered as a reason a man might commit suicide if he had lost, or was on the verge of losing, his land.

My informants utilized the notion of pride to convey appropriate or justifiable levels of self-esteem, self-respect, reputation, and even prestige. Crucially, it also signified independence—a point to which I will return later. Implicit in pride, then, were culturally defined and prescribed ideas about gender and role expectations. For instance, ideal characteristics of a farmer included being a rational and competent decision maker, a decent provider, and kind, generous, and good with children. Behavioral benchmarks included taking over the family farm, raising a family, passing on the farm to the next generation, and, in some cases, farm expansion. It is interesting to note, that definitions of "man" and "farmer" often became conflated—so closely linked were these concepts in the minds of my informants. Because of this, for the men and women with whom I worked, "farmer" generally meant male farmers, which is the way I use the term in this study.

My intent here is to tease out strands of cultural meaning that constitute the emotion of pride, to make intelligible the experience and social action (or inaction) of farmers who were under heavy financial strain. Before

I proceed, however, it is important to briefly review how emotions have been understood within anthropology to clarify my position and the cultural material presented in this study.

❦

Previous to 1980, emotions in anthropology had been studied primarily as phenomena that were largely psychological and biological in origin and were essentially universal, though they could respond to cross-cultural imperatives (Abu-Lughod and Lutz 1990:2).

Studies in psychological and psychiatric anthropology that reflect this traditional view illustrate the power and legacy of psychoanalysis, learning theory, and ethological and attachment perspectives. Their unifying principal assumption is that humans, cross-culturally, share the same stages of psychological development. Hence, particular situations can serve to activate or trigger universal feelings. For example, the birth of a child would arouse "natural" maternal love and mother-child bonding. Studies informed by these discourses view culture as functional, as serving to fulfill universal psychological human needs such as attachment or love. Finally, psychiatric and psychological anthropological researchers assumed that when normal stages of human development had not been realized or had been compromised in some fashion, humans cross-culturally would experience similar aberrances in emotional states that could be predicted and thus codified.

Even though Gilmore's (1990) *Manhood in the Making: Cultural Concepts of Masculinity*, does not treat emotions as its principal subject, it does serve as an example of a work that incorporates essential and universal ideas of human development to explain cultural phenomena. In his book, Gilmore seeks to understand the nature of masculinity and its manifestations in behavior and social life. He advances the psychoanalytic understanding of masculinity, allowing for the possibility that social learning can play a role in the formation of gender roles. In fact, his book provides a cross-cultural perspective on masculinity via descriptions of gender in particular societies around the world. Even so, Gilmore essentially views masculinity as: (1) the result of the necessary process by which males establish an independent identity from the mother, and (2) culture's solution in fighting the regressive tendency in men to re-merge psychically with the mother. In this way, cultures are able to produce independent men who will work toward the furthering of cultural goals and objectives.

Because the "true" nature of ideas that underlie gender and emotions may lie in the unconscious of the individual, a split occurs between behavior that can be observed objectively and ideas and feelings that cannot be accessed via standard anthropological research methods. This leads Gilmore, predictably, to speak of masculinity as primarily "performative": observable behavior that is the result of intrapsychic conflict to gain and continually prove one's independence. Thus, Gilmore adopts a largely functionalist perspective regarding the role that masculinity—and indeed emotions—plays in culture. Rather than seeing psychoanalysis as one of the many useful discourses in accounting for gender identity and social roles, Gilmore as well as Gregor (1985) and Brandes (1980) privilege universal structures of psychosexual development. In so doing these theorists reduce gender, and one could argue emotions as well, to pre-given psychosexual processes and close off questions of gender identity from history, the power of cultural expectations, social interaction, the realities and requirements of everyday life, and the possibility that individuals may have divergent cultural goals.

The result is that anthropological studies of psychological issues, curiously enough, were focused away from culture and toward individuals, or more accurately, toward universal structures of the mind and their local permutations in social groups. This perspective helped to cement another split—the split between emotions and culture. Since emotions were "centered" or universal, they were taken for granted and became more invisible to the social researcher.

Since 1980, emphasis in anthropology has shifted to the examination of the relationship between emotions and culture, specifically social behavior and relations. In doing so, anthropology has largely dispensed with universal ideas of psychology. Abu-Lughod and Lutz (1990:3–10) classify post-1980 studies into three categories: (1) analyses of the relativization of emotions which question universals by investigating, comparing, and contrasting conceptions of emotions in a variety of cultures; (2) historical analyses which examine changes over time in the manner a group of people think, talk, and use emotions in social relations; and (3) analyses which examine emotions in the context of social discourse, in the complexity of specific cultural contexts, and how they acquire their "meaning and force" from their "location" (Abu-Lughod and Lutz 1990:7). These later studies generally provide rich ethnographic descriptions of how emotions are understood, used by informants, and how they relate to broader cultural ideas.

They also frequently share a theoretical framework that has been informed by the work of Michel Foucault. Foucault offers us an understanding of personhood that is historically constructed and allows for the forging of connections "between forms of knowledge, institutional structures, and regimes of power" (Yanagisako and Delaney 1995:15) and, I would add, individual subjectivity. In his groundbreaking analysis, *The History of Sexuality* (1978), Foucault most clearly outlines his notion regarding the constitution of personhood. His examination of the historical and cultural development of ideas about sexuality and pleasure supports an understanding of subjects and their sense of self as both formed by and "dispersed in (multiple) texts, discursive formations, fragmentary readings and signifying practices" (Dirks, Eley, and Ortner 1994:12). He rejects depth models, such as psychoanalysis, of an essential inner self that is timeless and coherent. He views personhood, rather, as contingent and culturally and historically constructed in dissimilar ways in different times and places. The result, for Foucault, is diverging personhoods with infinite possibilities of emotional organization, different ways of knowing, and distinct configurations of gendered and sexual relations depending on the social discourses available to subjects.

I contend that in response to Gilmore, Foucault would argue that gender and identity are not performative, but that the discursive practices that help mold behavior also profoundly shape the emotional structure of the individual. Foucault, I think, would suggest that there are no distinctions to be drawn between practice and the interior life of the individual—both are predicated on available historical and cultural discourses for their construction.

Examples of this theoretical approach include the work of Abu-Lughod (1986), Lutz (1988), Scheper-Hughes (1992) and Yanagisako (2002). Their studies of Egypt, a Micronesian atoll, Brazil, and Italy, respectively, demonstrate that what emotions communicate can best be understood by examining their use and meaning from the standpoint of actors in particular positions of power within a social structure. Furthermore, they suggest a bond between emotions as gendered, learned experiences and beliefs and actions.

Scheper-Hughes's work in a Brazilian shantytown, for instance, examines the "naturalness" of maternal love and maternal bonding. The death of an infant in Bom Jesus da Mata may be a minor event, she explains. In fact, if a baby looks weak at birth or is undesirable in some way, mothers may practice selective neglect: they may discontinue feeding the infant and withhold attention until the baby finally dies. Scheper-Hughes argues that women's reaction to children's deaths in this context is a reasonable response to the unreasonable conditions of abhorrent poverty and dramatically high infant mortality rates. She illustrates that maternal love and bonding are not "natural" but the results of particular social-structural conditions and improvements in public health prevention strategies that help to ensure that most children in the developed world will survive infancy—creating sensible expectations among parents regarding the soundness of their emotional investments.

Studies such as those mentioned above also show that emotions exist within a system of power relations that serve a function in maintaining those relations. As Morgen suggests, this literature acknowledges that actors not only think but they also feel (1992:2), and it recognizes the political, economic, and ideological context within which emotions are generated and become meaningful.

For example, Lutz (1988:54) states, "In identifying emotion primarily with irrationality, subjectivity, the chaotic, and other negative characteristics, and in subsequently labeling women the emotional gender, cultural belief reinforced the ideological subordination of women." This book strongly supports that assertion, but takes it one step further—I argue that cultural belief also reinforces the ideological subordination of some men. Just as the Euro-American ideological construction of emotion promotes a certain vision of women, I contend that it also fosters a certain understanding of men and that this construction benefits different men differently.

On Oklahoma farms, men know how to talk about, and feel, pride. But to assume that pride results from—or that suicide or any other response of men to crisis is the consequence of—one discourse reflects a limited view of discourse theory. To take as a premise that men feel only pride is equally problematic. The danger lies, as Yanagisako and Delaney (1995:17) propose, in artificially circumscribing discursive domains to mirror our own categories as researchers, rather than those based on the lived realities of our informants: "While institutions and cultural domains of meaning have a profound impact on shaping ideas and practices, people do not necessarily organize their everyday actions according to these divisions. Rather, people think and act at the *intersections* of discourses" (1995:18).

Thus, our actions and beliefs are a consequence of a multiplicity of discourses. For example, the educational experiences of children, family relations and dynamics, friendships and romantic involvements, work experiences, and the various media all carry implicit and explicit assumptions, explanations, and values about emotions and, more generally, about subjectivities. Discourses are all around us. It is our socially constructed sense of self, mediated by cultural processes of power regarding relative advantage and oppression—our position within a social structure—that determines which ones we internalize, consider, act upon, and thus sustain or transform. Thus, we are continually remade or made new as we remake or make new the world.

I would argue that the statistics regarding farm suicides and family health substantiate that farm men do not feel less. Instead, hegemonic beliefs have advanced a certain ideal of masculinity and the male role that does not include the expectation that men learn how to give feelings of vulnerability form and voice. One broad consequence of this reality is the emergence or existence of communities that have never learned how to recognize the suffering and debilitation of men, much less how to respond to them.

My goal now is to examine cultural discourses that inform and construct the emotion of pride, which, in turn, informs the responses of my informants to severe financial distress. Below, I begin with four short essays that highlight particular aspects of northwestern Oklahoma culture. The first two are important because they draw attention to the historical and social processes that contribute to the construction of pride and have supported the cultural practices and subjectivities associated with it. The second two describe other processes that pose challenges to these notions of manhood and have served to devalue these same practices and subjectivities. All four essays are important, ultimately, in understanding how men were positioned within a matrix of cultural meaning to experience crisis as they did.

An Origin Story

Major portions of Oklahoma were opened to non-native settlement via land runs beginning in 1889. Stories of the Land Run of 1893, which opened the Cherokee Outlet, were important to my informants and were enthusiastically told to me. This is true not only because the event is relatively recent—non-native settlement is only approaching its fourth generation and some who made the Run, albeit as small children, were still alive at the time this research was initiated—but also because the Run and the struggle of those who staked a claim on the land loom large in the minds of my informants as origin stories. Yanagisako and Delaney (1995:2) state that, "[o]rigin stories are a prime locus for a society's notion of itself—its identity, its worldview, and social organization." It was apparent that stories about the Run were indeed an important societal nexus; additionally, they held great symbolic power and many of the men and women with whom I worked looked to them to help this researcher understand who they are, where they came from, and what they are like.

My informants' historical consciousness of this origin story begins with the Run itself and its immediate antecedents,[1] but the stories of this land really begin before, with the American Indian tribes that populated the region. Mrs. Perkins, who is herself of Native American ancestry, told me that if there is a theme that ties together the experience of both the historical inhabitants and the current farmers it would be that of the displacement of peoples.

As Euro-American expansion continued its forward march, Oklahoma was the "ultimate backwater of the rural-land frontier" (Morgan, England, and Humphreys 1991:xxi) and became the home for displaced American Indian tribes from what is now the continental United States. The push of the western frontier past the Mississippi and the expansionist desires of the nation caused a reexamination of the federal government's policies toward native tribes. The major interest of the government and would-be settlers was in the land—Mississippi, Georgia, and Alabama—held by the Creeks, Cherokees, Seminoles, Choctaws, and Chickasaws. Prior to 1830, the government had pursued a policy with tribes east of the Mississippi that was based on "civilizing" them and encouraging them to abandon the practice of communal land ownership in favor of individual ownership. The policy, to a large degree, was successful and—from the point of view of a government that was later in favor of the relocation of the tribes to the West—perhaps too successful, as the ability of Indians to contest white encroachment increased (Morgan, England, and Humphreys 1991:36–37).

As Euro-Americans' appetite for new land waxed and their patience began to wane, violence flared against the native tribes, who were determined to hold onto their land in the southeastern United States. Whites justified their increasing aggressiveness and encroachment of Indian lands as revenge, citing the violence perpetrated by Tecumseh and the Shawnee-Creek, who were British allies during the War of 1812. In addition, during this period state governments began to enact legislation intended to undermine tribal autonomy and power.[2]

Pressure was mounting in Washington for a political solution. The 1828 election of Andrew Jackson, an experienced frontiersman and Indian fighter, raised hopes that the problem would soon be addressed. Indeed,

President Andrew Jackson soon presented a proposal to Congress to re-move the tribes to Oklahoma. Gibson (1981:53) describes Jackson as being so obsessed "with driving the Indian tribes to the far frontiers of the United States that he gave his personal attention to the matter."

By 1830, Congress was ready to act. The hotly debated Indian Removal Act was passed by narrow majorities in both the Senate and House, and despite a series of challenges to the law that was made in federal court, ul-timately the Removal Act was implemented and the displacement of the southeastern tribes from their native home began.

In response to the law, a small group of Cherokees moved voluntarily. It was necessary, however, to forcibly remove the remainder of the tribe: "Under the watch of white militia, lawless whites drove the Cherokees off their land like cattle. Of the estimated eighteen thousand Cherokees who were forced to leave their land, approximately four thousand died in stock-ades or on the trail to Oklahoma" (Morgan, England, and Humphreys 1991:38).

> It is significant that most of the Indian removals [to Oklahoma] took place dur-ing [Jackson's] administration and that those not completed before he left office had been set in motion. The fulfillment of the Jackson removal program, with its ruthless uprooting and prodigal waste of Indian life and property to satisfy the president's desires and the demands of his constituency, has been aptly de-scribed as the "Trail of Tears."
>
> (Gibson 1981:53)

According to Oklahoma historians H. Wayne Morgan and Anne Hodges Morgan (1977:26), "no set of memories ever remained more vivid, or with more compelling reason, than did the tragedies of this march among the Cherokees."

By 1840, the removal of the five tribes to Oklahoma was complete and the area designated as Indian Territory was divided among them. Ironic-ally, many historians call the era between their forced relocation and the American Civil War the "golden years" of the five tribes. According to Morgan and his coauthors (1991:38), upon their arrival in Oklahoma, the tribes embarked upon a "truly remarkable experiment in Indian self-government" and they flourished. Each of the tribes created republican forms of governments with written constitutions and legal codes. Schools, including forms of higher education, were established. Newspapers were founded; books in English and native languages were translated and pub-lished. Towns became the center of commerce and, in rural areas, the system of plantation farming, which the tribes had practiced in the southeast, reappeared in Oklahoma, including their practice of owning black slaves. Because of their complex social organization, the Creeks, Cherokees, Seminoles, Choctaws, and Chickasaws soon became known as the "Five Civilized Tribes."

Their success attracted Euro-American interest in the region and whetted appetites for further expansion. The infiltration of Indian Territory began via commercial contacts and intermarriage, legitimizing the presence of whites. Their growing numbers in the area and their continuing desire for more land created pressure internally and externally for the opening of Indian Territory to white settlement. It was the native tribes' involvement with the Civil War that ultimately provided the federal government with the justification it needed to expropriate their land—land that had been promised to the tribes in the Removal Treaty for "as long as grass should grow and water run" (Debo 1940:152).

During the Civil War, the tribes were divided in their loyalties. Generally, the tribes whose land bordered the southern states sided with the Confederacy in the war. In addition to a common border, they also shared a sympathy for slavery. The diverging loyalties threw Indian Territory into violent chaos—halting the development of the tribes and leaving them significantly weakened and vulnerable to white expansion.

In 1865, at the close of the Civil War, representatives of the federal government met with leaders of the Five Civilized Tribes and several other Indian tribes. During the negotiations, the tribes were treated as defeated nations and, in 1866, were compelled to accept a settlement that included the abolition of slavery in Indian Territory and the forced relinquishment of lands in central and western Oklahoma, lands that were to be used for the resettlement of other tribes being brought to the region. As a result, the tribes would be removed to what is now the eastern half of Oklahoma.

During Reconstruction, white infiltration of Indian Territory increased. In addition, there was a new white influence with which to contend (Morgan, England, and Humphreys 1991:41). In the 1870s, cattlemen began occupying specific areas of the Cherokee Outlet—so named because it provided an opening that the Cherokee tribe could use to cross to their western hunting grounds. Tribal leaders in Tahlequah learned of the cattlemen's activities and sent representatives west to collect grazing fees. The fees were substantial and became an important source of revenue for the tribe (Gibson 1981:170).

With the quiet approval of the Cherokees, ranchers started to fence areas of the Strip. Other ranchers, who could not access the land, complained to the federal government. Officials finally forced the removal of the fencing (Bellmon 1993). To gain exclusive use of the Outlet and to protect their ranges from rustlers, a group of cattlemen established the Cherokee Strip Live Stock Association. They leased the land from the Cherokee Nation for $100,000 per year for five years. In 1888, at the end of the initial five-year period, they leased the land again for another five-year period, this time for $200,000 per year (Bellmon 1993). It is important to note that the Strip was crucial to cattlemen because, after the Civil War, Texas ranchers had to drive cattle through the area, on what has become known as the Chisholm Trail, to reach markets and transportation in Kansas to satisfy the demand for meat in the eastern United States.

In addition to the cattlemen, railroad companies were permitted to build on Indian lands as part of Reconstructionist treaties. These companies were the first to mount a major lobbying effort in Congress to open up Indian lands to non-native settlement. At first they were not successful; the companies, however, later capitalized on the increasing national need for cheap land and a growing feeling that communal land ownership, as practiced by the tribes, was evil. The fact that the Cherokees already leased large portions of their land to cattlemen did not help their cause (Morgan, England, and Humphreys 1991:42).

It was under this mounting pressure to open Indian lands that the government established the Cherokee Commission, which was authorized on 2 March 1889, to "negotiate with the Cherokee Indians and with other Indians owning or claiming lands lying west of the 96th degree of longitude in Indian Territory, for the cession to the United States of all their title, claim or interest of every kind and character in and to said lands" (Hill 1909:279). A proposition of cession was brought forth for consideration to the Cherokee chief J.B. Mayes and was rejected by him on the grounds that the Cherokee constitution forbade consideration of any such proposals (Hill 1909:279).

The commission worked with other tribes and brought about cessions which allowed the opening of other parts of Oklahoma to non-native settlement. Pressure to do the same was mounting on the Cherokees. The commission offered to buy the land from the tribe for $1.25 per acre. The Livestock Association, in sharp contrast, offered three dollars per acre.

In 1866 the Cherokee Tribe sent a delegation to Washington to meet with John W. Noble, the Secretary of Interior, to discuss their situation. Noble ruled that the Cherokees' title to the land was only an easement forfeited by their failure to use the land. Noble further ruled that the Indians' right to lease the land was void since they did not have a right to the land

3.1 Cherokee Delegation sent to Washington to negotiate the future of the Cherokee Outlet, 1866.

in the first place. In fact, he said, the government could confiscate the land, pay them nothing, and cut them off from their grazing income in the process. He advised the Cherokees to accept the government's offer of $1.25 per acre. Ultimately, the Cherokees received $1.40 an acre—nine million dollars less than they had been offered by the Livestock Association (Bellmon 1993).

After the acquisition of the land, the federal government decreed that the Cherokee Strip—composed of the present-day counties of Noble, Kay, Garfield, Grant, Woods, and Woodward—would be open for settlement via a land run. A date, 16 September 1893, was selected and announced across the country.

As time drew closer to the scheduled race, potential claimants began to gather in the areas bordering the Strip. Others illegally slipped into the territory beforehand. These "sooner" homesteaders would select a choice piece of land, hide in the brush, and at the right time after the start of the race, would emerge to stake a claim (Gibson 1981:181). The term "sooner" has become an important addition to the Oklahoma lexicon. The University of Oklahoma's sports team bears the name, as do countless businesses, organizations, and associations. Like the notion of *arranciarsi* in southern Italy or the *jíbaro* of Puerto Rican New York street life (Bourgois 1995), soonerism implies a certain craftiness and the possession of supreme survival skills.

Hill, in his exhaustive history of nascent Oklahoma, states that:

> There were charges that the town sites were occupied by sooners, acting in collusion with certain town companies and that the soldiers favored this occupation. There can be no doubt that the promises of speculation in the town sites attracted a large number of participants in the run. It is not disclaimed that the real homesteaders were in the minority in this and other openings. Thousands were here to get something which they then could sell to someone else at a profit quickly. It is claimed that the old pioneer spirit, willing to sacrifice and endure long years of toil in making the land productive, was an actuating motive to only a comparatively small number of these boomers and yet that small number remained to grow up with the country.
>
> (1909:300–301)

At stake in this great contest were 42,000 parcels of land, each 160 acres. Arkansas City, Honeywell, Caldwell, all on the northern border of the Strip (in Kansas), and Orlando, Hennessey, and Stillwater on the southern border (north of Oklahoma City), served as staging areas for the race. Thousands gathered at each of the sites. In all, between 100,000 and 150,000 people made the run while others waited in nearby towns to see if their loved ones had been successful.

At the time, newspaper stories about the Oklahoma Land Runs galvanized the attention of the nation like no event in history. One scholar has equated the interest focused on the Run of 1889, for example, to that of the first lunar walk (Lamar 1993:32).

According to Rainey (1933:277–279), the race was scheduled to begin at noon:

> Saturday, September 16, came on with a blazing sun and hot wind from the south. The dust rose in clouds, giving one some idea what storm-like clouds of dust would rise when the break was made. At five minutes to twelve all was tense and ready. The last details had been attended to and all were on the line. Horses champed their bits; locomotive safety valves were popping off and all eyes were on the man on horseback, stationed well within the Strip, who was to fire the signal shot which was to start the greatest race in the history of the world.
>
> Look out. A puff of blue smoke and the race is on. Everything helter skelter, pell mell, every fellow apparently for himself and the devil for the hindmost.

3.2 Contestants ready for the 16 September 1893 Land Run to claim 160 acres of free land in what is now northwestern Oklahoma.

Away they went the mass gradually thinning as the fleeter left the slower. The horsemen were, of course, soon in the lead, but all were doing their best. Many horses became exhausted and dropped from the race. Rigs were broken down and horses stumbled and fell, but the rolling mass moved on and within two hours the tides from the north and south had met near the middle of the Strip. The faster riders were able to secure the better claims, but many a straggler in a covered wagon managed to procure a home. Some of the wagon covers were adorned with humorous mottoes. On one was scrawled: "White capped in Injianny, chinch bugged in Illinoy, sicloned in Newbrasky, bald knobbed in Missoury, Oklahomy or Bust."

The chaos, noise and intensity of the race have been compared to that of thousands of cattle on the run. The chaos perhaps had to do with the odds, which were not in favor of the contestants: only about one in ten participants would stake a claim (Thompson 1986:49).

Men and women both participated in the race. Mrs. Nelson told me the humorous story of two relatives of Mr. Nelson, both young single women at the time, who made the Run. Both staked claims and, Mrs. Nelson added, "They both found husbands real fast!" A documentary video about the Run created for the centennial celebration (Bellmon 1993) describes one young woman who was decked out on a single horse that one could tell had been extensively trained for the event. The man telling the story

3.3 Start of the Cherokee Strip Land Run, noon, 16 September 1893.

said he started a conversation with this impressive young woman. The man said he told her that he would be the first to reach Wharton. She laughed and responded, "If you watch me, I'll show you the way!"

Not only did potential claimants have to cover significant distances, but contestants also had to ward one another off from claimed and wanted parcels. Claimants would have to stick up a flag on their parcel and then go to the nearest land office to file the claim. One man described how, after staking his claim, he watched the birth of the town of Perry in Noble County—from one individual to a camp city of over 10,000 in only six hours.

Once settlers arrived and staked their claim, their lives certainly were not easy. Droughts, severe thunderstorms, tornadoes, and prairie fires all threatened their ability to forge a profitable farm operation from the land they had won. Oklahoma historian John Thompson (1986:51) believes that the experience of homesteaders should not be romanticized, but, at the same time, "there is much truth to the popular image of these frontiersmen and women." Thompson writes:

> They were trying to make a living on land that received an average of 22 to 35 inches of rain a year. They had to farm in summer heat of well over 100 degrees, suffocating dust, fifty-mile-an-hour winds and subzero winter temperatures.

3.4 Enterprising sidewalk merchants sell their grocery items to prospective land owners waiting to file their claim at the land office located in Enid, Oklahoma, 16 September 1893.

Many homesteaders slept in sod huts which let in the weather so that they would awaken to find a layer of snow on them. Moreover because of the shortage of wood, they could never adequately warm their huts. Although sod huts did not adequately protect humans from the weather, rattlesnakes found them excellent shelters and would crawl underneath beds and into the rafters.

(1986:51)

It is estimated that up to fifty percent of the new settlers lost their land. This, however, was just the beginning of their struggles. The pioneers settled the region during a severe depression. As Hill (1909:301) notes:

The Cherokee country was opened at a period of profound financial and industrial depression throughout the nation. As a consequence, many of those who took part in the rush were actuated by hopes of finding here a country of free gifts and bounteous plenty where their distress would be quickly relieved. In the line that day in September were persons who had been thrown out of regular employment by the panic in other states. Many were entirely ignorant of the conditions that confronted them, had no conception of the hardships of a new country, were inexperienced as farmers and without means to support themselves through the pioneer period. It is not surprising, therefore, that after the boom and brief period of hope and exhilaration that followed the rush, many hundred yielded to their disappointment and left the country.

Later the region would be faced with the challenges of the Great Depression that gripped the nation and the Dust Bowl fast on its heels. The experience of the state during this era has become part of the American conscious, immortalized in works of social realism by artists such as John Steinbeck, Dorothea Lange, and Woody Guthrie. Informants told me their own stories of grinding poverty, during their own childhoods, and of their unending battle with the dust.

Despite the challenges, many stayed and it is those who stayed that concern us. The symbolic power of the stories of the Run lies in what they communicate about the people who are linked to that event: that they are tough, independent of mind and spirit, and that they can overcome great adversity. The stories tell us that despite the challenges of the Run, the profound struggles of the pioneer settlers, in spite of the Dust Bowl, Great Depression, tornadoes, and the many cyclical booms and busts—these people are crafty survivors and their continuing presence on the land is living proof of this fact.

3.5 Dust Bowl days in western Oklahoma, 1 October 1939.

The power of these stories, however, is not limited to the men and women with whom I worked or to descendants of families who made the Run; they hold great symbolic power for all of northwestern Oklahoma. This was made clear to me in an unexpected way. While I was conducting fieldwork, the feature film *Far and Away* was released in theatres across the country. The film, starring Tom Cruise and Nicole Kidman, depicts the life of a young Irishman down on his luck and marginalized by his community. He learns, through newspaper reports, that 160 acres of free land will be given away in Oklahoma Territory to the winning contenders in a great race. The film documents his struggles to get to America and Oklahoma, to the running of the race, and to his predictable success in staking a claim. I went to see the movie in Enid at a regular, evening showing. At the end of the film, I witnessed something I had never seen before in my hometown in a movie theatre: a cheering, standing ovation. It was clear that the film struck a chord with audience members, that this was *our* history that was being portrayed.

More indicative, however, of the resonance of this story is the fact that the opening of the Cherokee Strip is commemorated annually by communities across the region. In Enid, the largest town in the Strip, the celebration is huge and the focal point of community events, including rodeos, parades, craft shows, fairs, exhibits, and school reenactments.

As many of my informants were in danger of losing their land through foreclosure, their own stories stood in grim contrast to the nearly mythic narratives about the Run their ancestors had made and their success despite overwhelming obstacles.

In this respect, the stories of the Run also convey something that is critically important in understanding the cultural predicament of my informants: the importance of land not only for survival, but also as a central feature of their self-understanding. The life histories collected from the subjects of this research reveal that the Land Run of 1893 was often the staging area for their own accounts, the place from which they would start to communicate their long-term relationship with the land and with Oklahoma. Their stories led me to conclude that land was a complex and diversely constituted symbol. First and foremost land was, for my informants, about kinship.

Ghosts

The relationship between kinship and land was, for me, one of the more interesting aspects of this project. Farm families spoke about land in intimate ways, as if the land itself were a member of the family. For example, social gatherings that precede foreclosures on farmland are called "wakes"[3]—which always made me wonder who, exactly, had died? So compelling were my informants' observations regarding the bond between family and land, I began constructing kinship charts to see how and where I could place the land: Who would it be related to? What gender would it have? The nature of the relationship between families and their land became a mystery I wanted to solve, a central problematic I wanted to understand.

What is clear is that land is linked to the patrilineage. Kinship beliefs and practices support male identification with the land through the generations. Even though partible inheritance is practiced, the distribution of the patrimony is such that men always inherit land if it is to be transmitted and women may inherit land, but often inherit money instead. If both men and women inherit land, often the male sibling will provide monetary compensation to his sister for her parcel, particularly if he is interested in maintaining the integrity of the operation. Integrity here refers to the sum of the previous generation's land holdings to be transmitted to the next generation and, occasionally, it can also refer to land that was at one time part of the patrimony, but was not so at the time of transmittal.

A sister may sell her parcel because she may not have as strong an attachment (socially learned) to the land or because her spouse may not be a farmer and thus not interested in the land. If he is a farmer, his own land may not be conveniently located to the inherited land, making it difficult and undesirable to farm both properties.

Occasionally, not even all the male offspring will inherit land. At times it is acknowledged that a farm is simply not large enough to provide an adequate income for all the offspring; to divide the farmland ensures that none of the heirs will thrive economically. In this situation, only one male may be selected to inherit land, creating hard feelings among siblings—as it did for the family of one of my informants. Another option is to incorporate the farm and divide shares in stock among family members—as another informant's family had done.

A driving force behind inheritance practices is the desire for land to remain in the family and preferably in the family name.[4] Women cannot (or rather, did not, in my sample) contribute to this pattern because their names change when they marry (which they always did among research participants) and because there is the expectation that they will marry into another patrilineage. Mr. Gains, a family farm activist and the subject of chapter 5, stated, "There are some cases where, well, there wasn't no son. There are even cases where girls kept the land, but you know, it changed names, which is against what some of our forefathers wanted, but a lot of times that can't be helped. A lot of times there's nothing but girls."

That the bond between men and their natal land is very strong is not a point of contention. Farm couples became impassioned in describing farmers' relationship with the land. Biological metaphors were almost always used to convey an understanding of this bond. For instance, a farmer might say, "Well, I don't know if I can explain it—it's just that this red dirt flows through my veins." It also seemed that their biological tie to the land is what legitimized their ownership of land in their own eyes. For instance, Mr. Cobb, in making a clear distinction between his views about land ownership and those of a banker, said, "I think for a farmer, land is part of his lifeblood." He continued, "I worked that farm with blood, sweat, and tears; a piece of paper don't mean a thing." These comments reveal two things: that, like blood (see Schneider 1968), land symbolized kinship and that, like kinship, it is a relationship that encompasses long passages of time.

This kinship, this almost biological tie developed between a farmer and a particular piece of land—not just land as a general category. One man in the far western reaches of northwestern Oklahoma told me that since he was farming his wife's family's land, he did not feel at home there. He had lived on that land for over fifteen years, was raising his daughters there, and was very active in the local community. "This is not home," he said. "My home is fifty miles north of here where my father is still farming." To some degree he felt frustrated and told me he felt like a guest on his wife's family's land and was anxious to be farming his natal land, which he eventually would inherit.

I asked Mr. and Mrs. Thompson, neighbors of the Beckers, about the connection between kinship and land. I gave them a scenario: "Let's pretend you were going to lose your land. Is it harder to lose a piece of land that you bought or one that's been inherited?"

Mr. Thompson said, "It's gotta be," referring to inherited land.

Mrs. Thompson added, "I'm sure if we all of a sudden had to come up with a couple of hundred thousand, it'd be the land we bought more recently and that's what we would sell."

"This would be the last one," Mr. Thompson said, referring to the land we were on at that moment and where their home was located. "This is where my folks . . . this is what we call our homeplace . . . my folks' homeplace."

I asked them, "Why is that? I mean, why is one piece of ground more meaningful than another piece of ground?"

Mr. Thompson responded, "Well, it's just the way you been raised I guess."

Significantly, Mrs. Thompson added, "Just the memories connected with it."

Mr. Thompson continued, "I . . . when my dad gave us land, he never told us not to sell it, but we knew we better not. But he also told us that we could mortgage the land he gave us to buy more if . . . I don't know . . . maybe if we'd sold the farm, he'd probably kicked the lid off the coffin. I would hate . . . nothing you can do but I would hate to see . . ."

Mrs. Thompson jumped in, "Maybe it's a feeling of obligation that's owed to the ancestors to keep the land in the family."

"And I've heard that before," I said. "But how is it that the ancestors . . . how are they present in your lives? Does that question make sense? I mean, how is it that you feel that obligation?"

"Just do," Mr. Thompson said quickly.

Mrs. Thompson continued, "I guess because the memory is still here. His parents will have been dead twenty years [soon], but their faces and their speech, their mannerisms, and everything are still as real to me as when they were alive. Their spirit . . ."

". . . is still here," Mr. Thompson said, finishing her thought. "I still . . . they're still people to me."

Mr. Gains made several similar statements about this bond and the presence of the past and his ancestors. He said, "I feel like if there's a heaven and a hell you know, it's right here and I am inclined to believe that the spirits of my parents and my grandparents probably dwell in the place they loved. You know it seems to me . . . it makes sense at least and for that reason I feel like my granddad's spirit is still right there with that old land. He loved it, he raised his family and my dad and mother lived there for forty-five years. So I think if they have any choice when they get to that other side, that's where they'd like to be and that's probably where they are."

Country neighbors,
Flesh and skeleton,
Ancestral homestead
Adjacent to the modern place
Built of smooth-painted wood,
Red brick or aluminum siding,
All with storm windows
To keep the prairie out—

The living and the dead,
Time's compatible symbiots.
The old place doesn't light up at night—
Never was outfitted for natural gas or electricity;
It once ran on tenacity alone,
Still runs in the family
Like stubborn old blood.
They are not dead,
They who lived here;
The living give no thought
To tearing the old place down;
This old dried wood
Creaks with conversation;
The old house
Is no shrine to the dead;
It is still
A dwelling place.

—"Country Neighbors," Howard F. Stein

As I briefly mentioned, the historical relationship cements the bond between a family and its land, allowing it to be increasingly incorporated it into the kinship schema. This is true of both bought and inherited land as the following two examples demonstrate. The first example involves bought land.

In response to a statement they had made, I asked the Nelsons in what way land was a part of Mr. Nelson.

Mrs. Nelson responded, "It's his baby."

Even though I had by now realized that land did not fit on a kinship chart in any conventional way, my informants' responses revealed that one position it did resemble was that of an offspring. The Nelsons, as well as other farmers, spoke of land in very nurturing ways. Farming, for the male subjects of this research, was a generative and creative process. They said that, like children, land reflected the effort and time you put into it. Mr. Nelson said, "If we took on a new piece of ground, the way I always farmed it is I would go in there and plow it, start at the outside and plow to the center. Because everybody that had farmed it before I did had gone around and around the outside and thrown it out, so the land had a big ridge out here and there was a water hole out in the middle of the field. So I'd start plowing it and every year for twelve to fifteen years, why we'd be plowing it together, trying to get that soil back out there so you could level it up so the water could go ahead and drain off to the ditches. And so it was an ongoing process. It wasn't just today and tomorrow, or next week, it was years at a time."

Like Mr. Nelson, other farmers told me that creating a relationship with a piece of ground involved developing a long-term plan to get the land in the wanted shape, so that it becomes a good and productive farm. In addition to plowing strategies, the plan might also include fixing fences, completing terraces to prevent the runoff of topsoil, the removal of debris, adding extra fertilizer, liming, constructing watering holes, etc.

The amount of time spent and the accretion of experiences on inherited land also seemed to solidify the bond between families and their land. Previously, when referring to her husband's parents who had died, Mrs. Thompson noted that in her memory, "Their faces, their mannerisms, and everything are still as real to me as when they were alive." It was these small things, these unheralded everyday experiences that connected farm families to their ancestors. It was the daily experiences of working and living together on the land that bonded generations of the same family. Land reinforces this connection to kin because it serves—and is valued—as a mnemonic device.

Mr. and Mrs. Ross took me on a combination walking and riding tour of their property. Though they have struggled somewhat with financial indebtedness, this couple—particularly Mrs. Ross—has been extremely proactive in exploring alternative uses for their land, including organic farming. They were particularly proud of a few head of cattle they had recently purchased, from which they intended to produce organic beef. The cattle were exotic and certainly did not resemble anything I had ever seen before on an Oklahoma farm. As we were walking closer to them to try to feed them out of our hands, Mr. Ross began to talk about his family. As we continued our tour, I noticed that he was able to relate particular stories about his family, living and deceased, based upon our location on the farm. His favorite spot was an area where he and his father had worked cattle together for many years. It was a spot where the two shared many important conversations, where Mr. Ross received advice "about living" from his father. Mr. Ross turned to me at one point while we were still at that location and told me, "I still come out here and talk to my dad when I need to."

But often, and significantly, the bond created between land and families was less casual and much more deliberate. Social contracts or promises between generations of men to keep the land in the family made the presence of ancestors very real. In our continuing conversation about inheritance,

Mr. Gains said, "Like I told you before, that was one of the last requests my granddad made, and that was the same feeling my father had, that he wanted that old land to stay in the Gains family, so that's the only way of doing it. If we can somehow leave it to one of our sons, or all of our sons, why then, we'll provide then, another generation."

I would argue that the motivation for these social contracts has its origins, for many, in the stories about the Land Run and the overwhelming challenges that the state's early settlers faced, even as recently as the start of World War II. For both the early settlers and the subjects of this research, land was a form of familial and economic security. The feeling was and is that if worse comes to worst, the land could be used for subsistence farming, simply to support the survival of the family. In the very worst of circumstances, it could also be sold—however this was not even considered a possibility then or now. Farmers' tenacity and determination to hold onto land, then, can be understood in part as a reflection of the importance and the centrality of land in families' survival schemes.

My informants were quick to point out that their families, despite these historical challenges, stayed, endured, and even triumphed over hardships. This, indeed, is supported by historical evidence. The image of "Okie" families fleeing the Dust Bowl of northwestern Oklahoma and heading to a better life in California is, in fact, not an accurate image. In his comparative sectional analysis of Oklahoma, Thompson (1986:218) notes that "an estimated 97 percent of the net population loss" between 1930 and 1945, "occurred in the eastern half of the state, which was not crippled by dust and underwent almost no farm mechanization." The migrations from the east, instead, were due to a restructuring of the area's agricultural strategies that essentially replaced the cotton industry—and displaced thousands of cotton tenant farmers—with cattle farming. These displaced tenant farmers, by and large, constituted the famous migrations to California, with a large contingent of Oklahomans settling in the Bakersfield area.

Northwestern Oklahomans stayed in place and managed to survive and hold on to their land—a goal shared by the current generation of farmers. In working to achieve this goal, they believe they are honoring the legacy of their ancestors' hard work in building a farm operation. They also believe they are upholding the promise made to previous generations to keep the land and operation intact.

One day, I was staring at one of my kinship charts, still trying to figure out where "land" goes. My penciled hand began to move, and I drew a rectangle representing land around the entire kinship chart. This was it. Land encompasses kinship: it gives birth to it and it symbolizes it, and figuratively and often literally, it entombs it. Land, like "man" or "woman," for instance, is not a "fact" of nature that can be understood apart from its cultural meaning. The terms "farm family" and "farmland" cannot be understood independently; each relies upon the other for its meaning. It is only recently that the link between land and kinship has begun to loosen.

Trading Neighbors for Land: Community Change and the Incorporation of Technology

The increasing isolation of families in the rural communities of this study contributes to the experience of crisis of both farm men and women. Uniformly, my informants described substantial changes in rural culture during their lifetimes that have significantly lessened the social ties of community. These changes are inextricably bound to mechanization and technological innovation in the agricultural sector and to the increasing hegemony of the values of industrial agriculture, affecting both occupational and social practices.

A conventional measure of economic development is the proportion of the working population engaged in agriculture (Rogers 1991:52; Gupta 1998:38). There is an inverse relationship between the variables: as the percentage of individuals involved in agricultural production decreases, development is presumed to increase. Mechanization largely drives this decreased demand for labor. The following example is illustrative. According to the U.S. Department of Agriculture (Adams 1994:74), fifty to sixty person-hours per acre were required for wheat production in 1822 (using hand technologies except for a walking plow); but in 1890, only eight to ten person-hours were required. By 1930, with the use of tractors, combines, and trucks, merely three to four person-hours per acre were needed.

Technology can also propel what two of my informants, who were agricultural economists, called the "over-production trap," a phenomenon highlighted by Willard Cochrane (1993:427–29) in his "theory of the treadmill." The theory essentially describes the process of economic adjustment in the agricultural sector to a specific technological advancement. The adoption of a technological innovation, Cochrane argues, allows an individual farmer to outproduce competitors—for a while. As the technology is incorporated by the farmer into his operation, he has the capacity to produce more at a lower cost per unit. He will enjoy an advantage over competitors and realize more profit only until the particular technology is widely adopted by other producers. At that point more of the commodity

3.6 Scene on R. K. Wilson Farm in northwestern Oklahoma, 1910.

will be produced overall. As a result, the price of the commodity will begin to decrease and the farmer will get a lower return. In order for the farmer to maintain his economic quality of life, he has two options: (1) he can acquire more land, which would enable him to produce more, but it might also necessitate additional investment in mechanization and technology, such as a larger and more powerful tractor; or (2) he can get off the farm. Cochrane likened the effects of technology on producers to a steadily accelerating treadmill off the end of which farmers who are unwilling to invest in more efficient technologies must inevitably fall.

Many did fall off. Instead of choosing growth to keep ahead of the competition, many, historically, have chosen the second option and have left farming. Mr. Gains told me, "Man, I seen my neighbors leave by the dozens, that were my age. They didn't want to get big. They had milked cows and worked hard and they didn't want to get any bigger, so they just quit. The section [one square mile] I lived on, there was three families in that section. We have been the only ones left there for twenty-five, thirty years. [Once we had] seven neighbors. Today, my son lives over there on the old place, there's only three, so it took out four-sevenths of them. And everything just keeps getting bigger and bigger."

Mr. Gains' assertion that everything is "getting bigger" is supported by quantitative data. As technology has enabled farmers to work far larger amounts of land and to achieve greater yields, it has created an increased demand—and competition—for land. As farms expand in size, there will be fewer of them, assuming that little or no new land is brought into production. Strange (1988:176) graphically depicts this historical reality and shows that the number of farms within the United States fell from over six million in 1940 to less than two million in 1980. The graph below documents a similar pattern in the eight counties (Ellis, Garfield, Grant, Kay, Kingfisher, Major, Noble, and Woodward) in which I conducted fieldwork.

It is evident that there has been a substantial declination in the number of farms from 1959 to 1997, the date of the most recent agricultural census for which data is available. In 1959 there were 10,909 farms in the eight counties of interest to my study. By 1997, that number had decreased by 38 percent, to 6,722, a drop of 4,187 farms.

The trend toward increased farm size is also present in northwestern Oklahoma. Table 1 presents the average farm size for the same eight counties, over the last four decades. From the table and the graph that follows it, one can discern an overall pattern of growth.

The graph above shows that in 1959, the average farm consisted of 503 acres. By 1997, average farm size had increased to 698 acres, a 28 percent

Number of Farms, Eight-County Area, Northwestern Oklahoma, 1959–1997

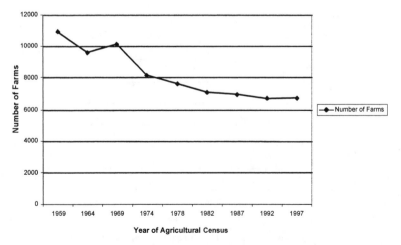

Source: U.S. Census of Agriculture, U.S. Department of Commerce, Bureau of the Census. Obtained from the Oklahoma Department of Agriculture, Food and Forestry, Statistics Division.

TABLE 1 **Average Farm Size (in Acres)**

COUNTY	1959	1964	1969	1974	1978	1982	1987	1992	1997
Ellis	853	978	957	1004	1036	1082	1072	1168	1077
Garfield	345	384	398	458	499	523	536	575	575
Grant	412	464	474	546	625	631	673	785	850
Kay	320	382	379	432	448	489	468	495	505
Kingfisher	375	426	440	503	530	537	507	567	556
Major	438	472	466	548	594	570	560	583	560
Noble	423	501	447	470	514	532	534	555	559
Woodward	860	913	850	948	971	975	912	921	902
Average:	503	565	551	614	652	667	658	706	698

Source: U.S. Census of Agriculture, U.S. Department of Commerce, Bureau of the Census. Obtained from the Oklahoma Department of Agriculture, Food and Forestry, Statistics Division.

Average Farm Size, Eight-County Area, Northwestern Oklahoma, 1959–1997

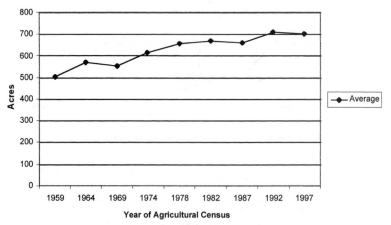

Source: U.S. Census of Agriculture, U.S. Department of Commerce, Bureau of the Census. Obtained from the Oklahoma Department of Agriculture, Food and Forestry, Statistics Division.

jump. An even more important indicator of the changes in the structure of agriculture and its impact on the participants of this research is the rapidly falling number of mid-sized farms, a topic I address in chapter 5.

According to Cochrane's treadmill theory, if a farmer is interested in staying in the profession, he has an economic imperative, or personal motivation, for expanding the size of his operation; a farmer would naturally want to increase the rate of return on his capital investments to improve his economic situation. But the incorporation of technology into rural areas and the increased competitiveness for land must also be understood in the context of industrial discourses, whose values have come to dominate the American agricultural sector.[5]

Farmers with a predominantly industrial orientation view farming as a business and not simply as "a way of life." These farmers are interested in organizing their operation to maximize productivity, efficiency, and profit. Technological innovation is a major concern of these producers. Financial risktaking and continual indebtedness is an expectation of business operation to fuel capital investment in technology, land, and continued growth. In this competitive environment, the quest becomes to substitute "my capital for your labor" (Strange 1988:169), that is, to replace as much labor as possible with technology, which is deemed more efficient.

The values of industrial agriculture have become largely institutionalized in the agricultural sector.[6] For example, the large land-grant universities, increasingly financed by major food-related corporations, gear their programs of teaching and research toward production and technological innovation. This is not surprising and, in fact, this is why they were created. However, the critical assumption underlying such programs is that profit—and the lifestyle it supports—is the ultimate goal, achievable if one makes the right management decisions.

One's skill as a farmer then becomes manifest through one's operation. My fieldwork indicated that, generally, the more land one has, the higher one's position in the local prestige system. Bigger—according to the logic of industrial agriculture—is better. Many farmers would disdainfully describe instances of the blatant competition between farmers for land. Mr. Cobb, never short of opinions, told me that if a farmer died, others would be "lined up before the body was room temperature" for a chance to buy or rent a piece of his land.

The land that we were on at that moment had come up for sale recently. Because it was adjoining a piece of his land, Mr. Melrose felt compelled to buy it. But he's not really sure

why he did it. He regrets that it came up for sale—not that he bought it. He said that he
would only have this one opportunity to buy it (land does not come up often for sale, es-
pecially adjoining land, it is more convenient if you can get your parcels as close together
as possible). Anyway, buying that piece of land, he admits, has gotten him into some fi-
nancial trouble.

Fieldnotes, 9 March 1992

The incorporation of technology—not just agricultural technology—and the growing prominence of industrial values, especially competition, have fundamentally changed the nature of rural living in the communities in which I worked, primarily by decreasing the opportunities for interfamily contact.

In rural Oklahoma, especially among the early settlers, cooperation and the exchange of labor were vital to survival. Thompson tells us that neighborliness was the "salient characteristic":

> Travelers always knew that, regardless of how impoverished a homesteader might be, his "latchstring was always out." According to the custom of the country, a stranger was always welcome to take what he needed even if his host was not at home, as long as he cleaned up after himself. Not only did survival depend upon cooperation, but life would have been unbearably drab without genuine companionship. Homesteaders were often racist and capable of intolerance, but during this pioneer period the force of prejudice was minimal.
>
> (1986:51)

But even later, during the childhood of my informants (primarily in the 1930s and 1940s), there was greater mutual cooperation than today. I asked the Beckers, for instance, whether families helped each other out on the farm during their youth.

Mrs. Becker responded, "Yes! Yes! When I was real little, we still threshed our wheat instead of combining it. You know, the thresher would come and the people in the community would do the threshing together. And I just can barely remember that. But it's not . . . you know, some of the women helped each other cook for the crew. And somebody . . . one of the uncles had a threshing machine and he took it from place to place and there would be a crew of farmers that would help each other out with that job. And then later, my dad was one of 'em. And there was probably, oh, six or eight in the community that built silos to put feed in for the cattle."

"Upright silos," Mr. Becker added.

"Yeah, upright silos," Mrs. Becker continued. "And so these people helped each other build these in the first place and then every year they helped each other fill the silo. And it was just that crew . . . But they would always fill the silo together in the fall with cattle feed. And of course, when I was little, they butchered together. They got together to butcher."

I asked them, "Did that make for different social relationships do you think? I mean, is that the case today where families help each other out?"

Mrs. Becker responded, "No. We don't work together like that now."

In effect, machinery has replaced human labor in both the sowing and in the multistep process of harvesting the crop. One farmer west of Enid told me that it was not cost-effective to even share resources with other farmers as had been practiced in the past. The machinery is so expensive, this farmer reasoned, that he would have to maximally use the equipment to realize an economy of scale—thus serving as a disincentive for lending machinery to other producers. Also, because of its expense, many would not likely want to risk loaning machinery for fear of malfunction or breakage.

Unlike today, farm couples described an active rural social life during their youth in the 1930s and 1940s. Not having access to many of the popular forms of entertainment available today or in urban settings, families took responsibility for creating their own fun: there were drama societies that would hold community presentations of theatrical works, church socials, much more frequent interfamily visiting, and regularly organized parlor games.

Boundaries between families were much more fluid during this period, I was told. Child rearing was highlighted for me as an example. The families in my study asserted that responsibility for children was, in many ways, more communal then. For instance, if a child were to misbehave, any number of adults could scold the child. Corporeal punishment too, in schools, was an accepted form of discipline. My informants said that that type of communal responsibility for youth behavior is unimaginable today—so entrenched has our society become in legal discourses, including individual rights. "You can't do that today . . . you would be afraid of being sued," one farmer said to me in disgust.

The type of rural life experienced by my informants in their youth stands in stark contrast to the realities of rural life today, which can hardly be described as communal. When I asked my informants what they believed had been most responsible for the historic changes in the nature of social relations, I was surprised by the consistency of the responses. Uniformly, the subjects of this study told me that the introduction of the television was most responsible.

Their answer reminded me of the Barry Levinson film *Avalon*, which recounts the generational changes of a single immigrant family during the course of the twentieth century. The film depicts the intense interfamily social relations, the many opportunities for large gatherings, and the more intimate day-to-day contact that were at the heart of one's identity as part of the group. One scene is etched in my memory. In sharp contrast to the large social gatherings, which had been the rule, the particular scene shows a family, after the introduction of the television, having dinner. Only one nuclear family is depicted in this scene: each member has an individual food tray and is staring blankly at the television screen—not talking to one another. They are in their own house and not out in the community. This, in essence, is exactly what my informants described. There was no need to create "fun" since it was readily provided at home. As a result, the drama societies began to founder and organized parlor games and regular inter-family visiting declined in frequency and finally ended. My informants are convinced that television significantly contributed to the lessening of com-munal ties.

Farm couples would describe their early lives in rural areas with a cer-tain degree of nostalgia for community lost. Part of what has been lost is simply people. I was told that there was a time when there were "lots of people out here." There were more homesteads per section of land than at present and more daily interaction with other families. During their youth, the population of the region warranted schoolhouses, of the classic one-room variety, every few miles, the decaying remnants of which were evi-dent on my trips to interview families and conduct participant-observation research.

In addition to the displacement of labor caused historically by the in-corporation of technology, the farm crisis of the early eighties exacerbated this migration of people out of northwestern Oklahoma. The population of the northwestern area of the state makes up about ten percent of the total population of Oklahoma, of which seventy percent lives along the Tulsa-Oklahoma City-Lawton axis. Although the 1990 Census indicated that the population of Oklahoma grew 3.8 percent during the 1980s, this was not the case for the northwest. As jobs were lost due to failed busi-nesses and farms, the area experienced a 6.7 percent decline in population. Losing between 16 and 24 percent of their population were: Harper, El-lis, Woods, and Blaine counties. Alfalfa, Beaver, Grant, Roger Mills, and Garfield counties experienced population losses of between 10 and 15 percent (Rojas-Smith, Ramírez, and Perry: 1992:10).

The impact of this outward migration was most keenly felt in the eco-nomic sector that depended on a young, stable and well-educated work force to maintain its vigor. Because farms can usually only support one family, it was not possible for many young families to stay at home and farm. Without inheriting the farm enterprise, it was practically impossible for a young person to start his own operation due to the prohibitively high cost of farmland and machinery. The young would have to wait to inherit. Meanwhile there were few other occupational opportunities available. Migrating to urban centers was the choice for many.

> Bobby told Lucy, "The world
> ain't round . . .
> Drops off sharp at the edge of town
> Lucy you know the world must
> be flat
> 'Cause when people leave town, they
> never come back"
> They go ninety miles an hour
> to the city limits sign
> Put the pedal to the metal 'fore they
> change their mind
>
> —"Small Town Saturday Night," by Pat Alger and Hank Devito

There was considerable ambivalence in parents' feelings concerning the migration of their children. On the one hand, they lamented the loss of their young and realized that their goal of passing on their operations to future generations would not become a reality. On the other hand, considering the financial trouble that many of them experienced, parents did not necessarily want their children to pursue farming as an occupation. They accepted migration as an unfortunate, but sometimes necessary, reality that was in the best interest of their children.

We finished mending the fence. We climbed back into the pickup and, as we were pulling out of the homeplace to feed cattle, Mr. Melrose told me a joke. He said, "A farmer willed his son a farm and do you know what the son did? He sued his father for child abuse!" He said later that the worst thing that could happen to a young man is to inherit land, because he could be in trouble in a hurry if he's not a good manager and if he is not willing to start out slow.

Fieldnotes, 13 February 1992

As younger families left for the larger urban areas of the state, the tax base for basic social services and education diminished and set off a downward spiral of economic and social deterioration in communities throughout the region. Fewer children and fewer tax dollars resulted in school closings and mergers. Many of the established networks of clubs and organizations, the primary conduits of community life, have broken down due to a lack of individuals willing and able to commit time and energy to them. Those who remained in rural areas, increasingly the elderly, have been left with reduced fiscal and human resources to meet their economic and social needs (Rojas-Smith, Ramírez, and Perry 1991:11).

Mr. Gains told me, "Well, see, just gradually . . . see it's a little old town. It's been a ghost town as long as I remember, but still we had a shop and a little store, filling station, but there's nothing left up there but old people. And our school hurts because there's no young people with young families. We got down here at one point where average daily attendance was set at a certain figure, you know, and my gosh, well, we barely had enough kids to go to school. The other side of the district had a lot more people because it's closer to [town], but the valuation's so low that they couldn't maintain their school and buy buses and keep the school open. So we went together. And it's a good deal, but I don't know how long it's gonna last since they passed 1017 [educational reform legislation]. I think there's a few years and it'll be gone. Well, all those things add to the fact that it's going to just disappear."

This out-migration also was responsible for the belief expressed by many farmers that rural life as they had known it would soon come to an end. Many felt that they were, if not the last, among the last generations to have the benefit of the lifestyle that farming afforded—freedom and a connection to the natural world and to a moral good they felt was inherent to farming.

It is ironic that the agricultural practices and values endorsed by industrial agriculture are the very ones that propel negative rural social change and bring many communities to crisis—a process, by the way, that is not new (Fitzgerald 2003:121). Many of the men and women with whom I worked were conscious of their paradoxical position: they valued the freedom promised by modern capitalism, but they also "long[ed] for the security it destroys" (Dudley 2003:191).

The Political Economy of the Crisis

The seeds of the 1980s rural crisis and its ongoing aftermath were sown during the 1970s, a period during which the American agricultural sector was being increasingly incorporated into larger national and international economic processes. There were various reasons why U.S. involvement in international agricultural markets increased at that particular time. One influential factor was the dramatic expansion in the number and size of markets for American agricultural products. In 1972, an accord was struck with the Soviet Union in which that country agreed to purchase eighteen million tons of cereals from the United States (Schertz 1979, cited in Barlett 1993:169). As a result, one-fifth of the total U.S. wheat supply was sold and prices more than doubled. Experts agreed that the Soviets would continue to be large consumers of American grain well into the future. Additionally, President Nixon's shift in policy toward China signaled the potential opening of that awesome market to domestic farm goods. Finally, droughts in parts of the developing world translated into low world grain reserves and triggered concerns about global food security. As a result, agriculture became a strategic issue due to the world food crisis of 1972 and 1973 (Barnett 2003:163). In response to these conditions and the declining value of the dollar, many nations increased their consumption of American agricultural goods. Some were enabled by enormous loans from international lending agencies (such as the World Bank and the International Monetary Fund) that included credits for the purchase of specific commodities, including agricultural products (Adams 1994:234).

The other factors prompting increased American involvement in larger economic spheres had to do with U.S. tax and agricultural policies. The U.S. tax code of the early 1970s contained provisions that strongly encouraged investment. These included tax credits, rapid depreciation schemes and, especially, an income tax deduction for interest payments. At the same time, farm policies supported a panoply of price support and supply control programs which were "used to stabilize and maintain the price of selected agricultural commodities at artificially high levels" (Barnett 2003:162–163). The high returns on farming made possible through these policies encouraged "massive investment in agricultural assets" (Barnett 2003:164). In addition, loans made to farmers at below market rates, primarily for land purchases, helped to ensure that American farmers were well positioned to respond to the increased demand for food abroad. U.S. producers, in fact, demonstrated an impressive capacity to gear up production "to feed the world." Between 1970 and 1973, wheat exports doubled, from 19 million metric tons to 38 million metric tons (Barnett 2003:164), inaugurating a boom in the American agricultural sector.

My informants spoke about this period, from the early 1970s to the early 1980s, as a time "when farming was fun." The boom in agriculture meant that farmers could achieve measured progress—measured by the degree of expansion of their operations: the purchase of land and larger machinery and the upgrading of buildings and homes. And expand they did. Barnett (2003:164) notes that between 1969 and 1978, the value of American farmland and buildings climbed 73 percent, or $381 billion (measured in constant 1982 dollars). This improvement in the financial condition of farm families also meant that, for the first time, they could live at a standard similar to that of their middle-class urban counterparts. As a result, farmers' morale, sense of purpose, and of the importance of agriculture all increased (Friedberger 1988:7).

Friedberger (1988:6–7) asserts that, at the time, most farmers felt comfortable with philosophies and policies that focused on the maximization of production to meet the growing demand for American food abroad. In fact, agricultural experts aided this effort by promulgating technical advice that challenged producers to strive for maximum efficiency in their operations. A critical tactic offered to help producers achieve this goal was expansion of farming operations to achieve economies of scale. Farmers, in essence were told to "get big" or "get out."

Mr. Becker said, "A lot of people thought that things couldn't help but get better for the farmer. In the late '70s that was the idea, you know, everything is going to get better. The land will just keep going up. I know our son-in-law, you know that just got started farming at that time and that's what the bankers told him. 'Better buy it now because it's not going to do anything but go up.' But he's lost it because that was not true. It didn't happen that way. That was the general advice from bankers and the lenders as a whole . . . they advised farmers to do that."

As Mr. Becker's statement indicates, lending institutions were eager to provide loans to farmers interested in expansion during the boom years. The principal reason for this was that land—farmers' principal form of collateral—continued to surge dramatically in price. For example, in 1970, nationally, an acre of farm land sold for $419; by 1980, this cost had increased to $2,066 (Stock 1996:156). There were several factors involved in this increase. One was the high returns promised by agriculture, already mentioned. This naturally fueled speculation and competition for farmers' principal means of production. The inflationary economy was also a factor in increasing land costs. Rising prices became a national concern during this period, which prompted farmers to invest in expansion sooner to avoid paying higher prices later.

But a shift in values associated with agricultural indebtedness also occurred during this period that may have contributed to the problem.

Previously, farmers had looked upon farm debt as a necessary evil to be avoided if at all possible. However, the continuously increasing number of overseas markets, the prosperity experienced by many producers, and the increasing hegemony of industrial agriculture—which viewed ongoing indebtedness as an accepted and even expected business practice—encouraged farmers to leverage their assets as much as possible (Adams 1994:234).

The double-digit inflation the nation was experiencing—propelled by the actions of OPEC and consequent spiraling fuel prices—made double-digit interest rates seem inexpensive when loans were secured. For example, during the 1970s the real cost of borrowing—"the nominal rate paid by the borrower minus the rate of inflation—was negligible and, at times, almost 'free'" (Dudley 2000:23). As a result, farm debt soared. By 1978, national farm debt had reached $120 billion (Schertz 1979, cited in Barlett 1993:169).

In the mid- to late 1970s the price of agricultural commodities had begun to level off. As a result, farm incomes did not keep pace with the increasing costs of capital expansion and producers began to experience cash shortfalls. The gap between income and the cost of farm operation fueled further equity financing. Bankers were still eager to continue to lend to farmers because they were firm in their conviction—as Mr. Becker's statement indicated—that the price of their collateral, principally land, would continue to increase in value. At the same time and for the same reason, insurance companies and the Farm Credit System adopted new "aggressive lending practices" (Dudley 2000:24).

Dudley (2000:29–30) claims that if there is a master narrative about the farm crisis, it is this: the "march of progress and the triumph of technology over the limitations of human labor." It is a Darwinian storyline that extolls the virtues of those who have been able to survive the fierce competition that has periodically served to thin the ranks of American producers during the course of the twentieth century. The reality, however, is much more pedestrian. By the early 1980s, the constellation of factors that prompted dramatic agricultural expansion—principally global demand for American agricultural products and low-cost credit (Dudley 2000:33)—began to disintegrate. One of the principal causes of this dissolution was the failure of agricultural economists[7] and policymakers to appreciate the extent of America's involvement and integration into the world market economy. Policymakers were still convinced that the worsening international food shortages would continue to engender unlimited demand for U.S. farm products (Friedberger 1988:9).

Many of my informants reserved harsh criticism for President Carter and his decision to impose a grain embargo on the Soviet Union—a retaliatory move in response to the Soviet invasion and occupation of Afghanistan. My informants viewed this embargo as the primary cause of the collapse in agriculture. Although farmers were hit hard by the embargo—which did lead to a temporary market crash—it was the structural changes implemented by the Reagan administration that had more profound consequences for producers (Adams 1994:236).

Adams, in her historical study of the transformation of rural life in southern Illinois, discusses the fiscal and economic strategies pursued by the Reagan administration that had a devastating impact on American farmers. She explains that the administration sought to curb inflation and strengthen the U.S. dollar in international markets. To accomplish this, Reagan tightened the money supply through the Federal Reserve System, which caused interests rates to increase. The prime interest rate reached 20 percent in 1980—which made credit, which had previously been widely available, much more inaccessible. The administration also instituted tax cuts, which largely benefited corporations and the wealthy. Since the government now had fewer dollars and the administration was, at the same time, pursuing the largest peacetime expansion of the military in history, the federal government was in urgent need of money. It borrowed billions by selling high interest rate bonds to creditors around the world. The global demand for dollars to invest in U.S. Treasury bonds contributed to the soaring value of the dollar which, in turn, had dire consequences for the agricultural sector. Its products were now becoming too expensive for overseas buyers. The impact of this strategy on agricultural exports was profound: peaking in 1981 at $46 billion, by 1986 the exportation of food products had fallen 50 percent (Barnett 2003:168). In addition, the strong dollar helped to boost interest rates even further, "catching farmers in the worst cost-price squeeze [the balance between costs farmers paid and prices they received for their products] in history" (Adams 1994:236).

Simultaneously, the International Monetary Fund was instituting a series of changes in its dealings with developing nations, requiring them to adopt austerity measures to meet their repayment schedules. To comply and to raise needed revenue, many developing countries became much more aggressive in marketing their own agricultural products, creating increased competition for U.S. farmers. They also were unable to purchase as much American grain and other commodities as they had in the 1970s, further decreasing the demand for U.S. agricultural exports (Adams 1994:236).

Federal farm policy also played a role in the downtrend in agriculture. Overproduction and increased international competition were the results of the 1981 Farm Bill, which set prices above market values (Barlett 1993:169–170). As exports and crop prices fell, production costs continued to climb. This combination caused farm incomes to plummet. Land values also began to fall, threatening the level of assets bankers held as collateral on loans. Because of this, credit was cut off to many producers, forcing many out of farming and off their land. Nationwide, the percentage of farms going out of business nearly tripled from 1982 to 1986 (Murdock and Liestriz 1988, cited in Barlett 1993:170).

The Farmers Home Administration (FmHA) responded to the general crisis in agriculture by making emergency loans to farmers who lacked access to credit or who had lost crops. Barlett (1993:181–183) documents the excessive generosity of the FmHA. My informants too, often spoke angrily of FmHA's habit of almost forcing farmers to accept credit and prompting management changes within operations as requirements for borrowing.

For example Mr. Cobb said, "What makes me mad is that some of these guys that are ag [agricultural] lenders in the banks are just kids that are still wet behind the ears that get out of college and think they know more about it than the guy who's been out here for thirty years doing it. And they come out here and try to tell you how to do it . . . if that don't burn me! That just makes a farmer madder than hell!"

In addition, many of my informants also found these loans difficult to repay. Since they considered lenders experts in financial matters, some farmers could not understand why they would be loaned so much money given the reality of their financial situation. Many farm men and women with whom I worked attributed sinister motives to FmHA and to other lenders, some seeing it as a deliberate attempt to reduce the number of producers in an area. As Mr. Cobb's statement indicates, others saw it simply as bad management attributable to inexperienced lenders. Still others reacted strongly to the rapidly changing personnel in lenders' offices, particularly FmHA, which they saw as a calculated policy to reduce the personal ties between lenders and borrowers, making it easier for lenders to foreclose on farmland.

Mr. Cobb continuing his point from above: "It's not so much the regular banks, but it is that way in the PCA and the land banks and the FmHA. Now [in] those, those kinds of institutions, that is pretty prevalent. And then those outfits are big enough now that they've got problems in the country, in all three of those, that they move people around and essentially

what they're doing is making hatchet men out of them and they're bringing people into the community that don't know nobody and they look at
the numbers and if it ain't right, they shut 'em down and they don't care.
Where if they grew up there, it'd be awfully difficult to do."

Whatever the cause, the result was clear: in 1985 the Farm Credit System lost $2.7 billion, "the largest one-year loss of any financial institution
in U.S. history" (Barnett 2003:168); and by 1986 more than 75 percent of
FmHA's total loan portfolio was nonperforming or delinquent (Barlett
1993).

Dramatic expansions in the American agricultural sector had occurred,
almost cyclically, in the 1910s, 1940s, and in the 1970s. These short periods of prosperity, however, seem to create the illusion among members of
the agricultural community that they would be long lasting (Murdock and
Liestriz 1988, cited in Barlett 1993:170). Inevitably the conditions change,
land values deflate, and "an extensive period of depressed farm conditions
generally follows" (Murdock and Liestriz 1988, cited in Barlett 170). The
downturn in the agricultural sector of the 1980s, however, was different
from those of the past. This time farmers were more vulnerable due to the
changes in agriculture: their farms were larger, more dependent on expensive purchased capital inputs, and more indebted. Thus when the crisis
struck, as the Nelsons' story demonstrates, it was much more difficult to
make the necessary changes, such as cutting costs and substituting family
labor, than had been possible in the past (Barlett 1993:170–171).

Summary

I have argued that far from being simply psychological or biological
phenomena, emotions are largely informed by culture. Individuals are positioned within a particular matrix of power relations and discourses that
condition their understanding and experience of emotions. I have outlined
some important components of pride, as experienced by men in northwestern Oklahoma, including its link to cultural ideas about kinship, history, land, and inheritance practices.

The values of the story of settlement and of the hardship of ancestors have been transmitted through generations of families. The critical
lesson of this narrative has been that land means security and that it is the
familial—patrilineal—vault of history, as men become tied to their natal
lands through inheritance patterns that favor them in its transmittal. The

goal of maintaining land within the family is a prime objective of the families with whom I worked. The effort required to achieve this goal honors ancestors' struggles to acquire land, develop an operation, and maintain its integrity; it also honors cross-generational promises made to do the same. Holding onto land demonstrates a commitment to the past, to the ancestors in their midst.

Particular international and national political economic factors converged in the early 1980s that created monumental obstacles for farmers in their effort to keep their operations together. Part of the emotional trauma for farm men in financial crisis, created by the collapse of the agricultural economy, is the fear that they will not be able to fulfill their promise to keep family land. In addition, they feel guilty and incompetent: their ancestors had to go through substantial struggles to acquire, maintain, and even expand their operations. Supposedly, this current generation has all the amenities and technology it needs to succeed—and with no seemingly comparable challenges like the Great Depression or the Dust Bowl—and, still, its members are having substantial difficulties. Historically, the Land Run, the Depression, and the Dust Bowl were events that the entire community experienced; due to the overwhelming environmental and economic challenges, everyone was having a difficult time making their operations successful. The present economic crisis, they perceived, happened only to a few, adding to the shame and guilt many men experienced.

The hegemonic values of industrial agriculture and the expanding mechanization and technological innovations that have been historically incorporated into rural areas contribute to the loosening of social ties by decreasing the need for the exchange of agricultural labor and by undermining the motivation of social gatherings. Industrial values contribute to: (1) increased competitiveness among neighbors, (2) the already strong sense of individualism present in northwestern Oklahoma, (3) the feeling of individual failure, and (4) the community stigma surrounding farm foreclosures and bankruptcies. Finally, the lessening of the social ties of community has increased the isolation of nuclear families. These historical changes in rural communities have placed men in relative isolation when crisis strikes and with few community resources to address their needs. Conversant in a language of pride and much less so in a language of vulnerability, men, with few perceived options at their disposal, simply withdraw from communities and their families and just try to hang on as long as possible the best way they know how. As these statements suggest, emotions are not merely felt, but rather they are embodied thoughts that are necessarily evaluative of one's particular social, political, and economic context.

In the following chapter, I address other kinds of positionings critical to understanding pride and the experience of crisis among farm men: gender, farm management styles, and occupational role evaluation.

We might as well have a catechism:
What is a farmer?
A farmer is a man who feeds the world.
What is a farmer's first duty?
To grow more food.
What is a farmer's second duty?
To buy more land.
What are the signs of a good farm?
Clean fields, neatly painted buildings, breakfast at
    six, no debts, no standing water.
How will you know a good farmer when you meet
    him?
He will not ask you for any favors.

—Jane Smiley, *A Thousand Acres*

Are there not discursive conditions for the
articulation of any "we?"

—Judith Butler, *The Psychic Life of Power*

FOUR

The Good Farmer
Gender and Occupational Role Evaluation

MR. MELROSE and I were riding around in his pickup feeding hay to his cattle which graze on several different farms he owns. On the way back to his homeplace, he came to a slow stop, rolled down his window, and examined his neighbor's property. He shook his head in disgust and drove on.

I asked him what he saw. He answered that the farmer was still using the same old rickety machinery that he had been using for fifteen years.

"Is that bad?" I asked.

He answered, "A farmer has got to keep up to get ahead. Him? He's too conservative. He never borrows any money to improve his operation or expand. At this rate he'll be an old man before he realizes any profit from his operation. Then he'll be too old to enjoy it."

"Is he a bad farmer?" I asked.

"Well, he's not a good one."

To a large degree, Mr. Melrose's responses to my questions reflect the values of industrial agriculture, which I briefly outlined in the last chapter. His comments reveal a concern with expansion—bigger is better. They also indicate that debt is an expectation of business operation; the leveraging of one's operation enables its growth. Rather than labor, investments are made in technology and land in an effort to realize peak efficiency and maximize productivity and profits.

Perhaps more importantly, farming is viewed as a way to make a living; it is a business operation and not necessarily a way of life. Organization around the values of efficiency and profit maximization often precludes the involvement of family members, since technology is employed to replace labor. As a result, there is a marked separation between the private and public spheres. The focus is not on people, but rather, as Strange (1988:37) points out, "on an abstraction: management." An industrial orientation supports the view of farmers as rational, independent, and free agents whose decision-making skills are supreme in success or failure. This is a critical point.

As I have previously mentioned, the values of industrialization have come to prominence in rural areas. To a large degree, they inform the reckoning of prestige in farming communities. Individuals who follow its tenets are recognized as "progressive": "It is these farmers who get admirable stories written about them in the farm magazines, who win awards from small-town chambers of commerce (because they do a lot of business on main street), and who are considered top managers by agricultural experts," asserts Strange (1988:39).

In many ways the values highlighted above are reminiscent of those proposed by the Spindlers (1983:58) as characterizing the American national character: individualism, an achievement orientation, and the belief in equality to succeed.

During fieldwork, I learned that other philosophies and management styles existed among my informants, challenging my preconceived notion of farmers as a homogenous group. I had the opportunity to interview the farmer who was the subject of Mr. Melrose's comments. I, in turn, asked Mr. Ross if he thought Mr. Melrose was a good farmer.

He said, "Well, he's a good farmer if you mean he's big. Pretty soon he's going to farm the whole county and what for? I mean you can't take it with you. I guess we'll know that we have been successful if we can hold on to this place long enough to pass it on to our kids."

Mr. Ross's comments are in alignment with a more family-centered understanding of farming. Mr. Ross and others like him would dispute the notion of farming as a business or occupation like any other; farming, instead, is viewed as a way of life. No clear distinction is consequently drawn between the public and private spheres. Farm and family are mutually dependent: the family is an integral part of the production process—principally through its contribution of labor—and the farm serves an important role in the private sphere, through its organization of family life. Technology is certainly employed, but its use, ideally, is tempered to the needs of the family, to ease the labor burden and not necessarily to achieve "peak efficiency." The family is affirmed as the appropriate locus for farming and a certain moral good is thought to be intrinsic to the work and the lifestyle it enables. The family and its maintenance on the land are critical goals of family-focused farmers. Land is seen as a generational trust.

In their 1983 review article, the Spindlers assert that classic analyses of American culture by sociologists and historians have assumed—because of American men's traditionally dominant status—that the "characteristics of American men were the characteristics of American people, including women." Feminist scholars have effectively challenged this supposition by illustrating the diversity of women's experiences, their resistance to American patriarchal systems of meaning, and the generation of competing discourses.

Here, I continue to question this premise in the gender literature on American culture by demonstrating that the values of American men—often assumed to be held by all Americans—are not even applicable to all men. This chapter analyzes diverging farm management discourses among northwestern Oklahoma farmers and their implications for ideas about gender and gender roles, including occupational role evaluation. I seek to illustrate the diversity of men's experiences and perspectives within farm communities and thus challenge what Yanagisako and Collier (1987:26–27) have claimed to be an assumption of much of the feminist literature: that of a "unitary men's point of view." I begin by outlining ideas about gender that underlie family and industrial farm goals, strategies, and actions.

It is important to reiterate that all of my informants were family farmers—referring to the basic unit of production. Thus when I refer to industrially oriented farmers, I am still referring to family farmers and not to large conglomerates or their agents. In fact, due to factors in the state's early history, which I discuss in the next chapter, corporations have had limited success in Oklahoma's agricultural sector. This chapter, then, is about discourses of meaning competing for dominance among northwestern Oklahoma farm families.

Farming continues to be a predominately male tradition in northwestern Oklahoma. It is clear that this gender patterning is deep and institutionalized. Previously, I have illustrated the ways that beliefs about kinship support male identification with land through the generations, including the desire of land to remain in the family name and inheritance practices that favor men in the transmittal of farms. Thus an important relationship between men and their natal land becomes established.

This strong bond between men and land is also supported by other institutions and organizations central to farm life. For example, just prior to the initiation of my fieldwork, the policy of federal subsidy programs (extremely important for the financial viability of farms and for the promotion of economic stability within the agricultural sector) allowed only one individual per household to qualify for government payments. To be eligible, an individual had to spend most of his time in management, labor, and have a financial investment in the operation. This meant men. Mrs. Becker told me, "Until recently, [at] the ASCS office, as far as getting a government payment in farming, a wife, even though she owned as much as her husband and did as much work as her husband, still was not recognized as a person, as an individual. My neighbor works on the farm as much as her husband and that was really a sore point with her, that she was not even recognized as a person by the government."

The link between men and farming and the marginality of women is also mirrored in the structure of mainstream farm organizations such as the Wheat Growers Association and bedrock rural institutions like the Cooperative Extension Service. In the Wheat Growers Association, for example, it is primarily men that belong to the main arm of the organization that conducts business, provides educational opportunities, and makes decisions. Women belong to the auxiliary organization, the "Wheathearts," that my informants seemed to indicate, engaged in supportive kinds of activities ("How to make dumb baskets," said Mr. Cobb).

More significant in the daily lives of my informants was the Cooperative Extension Service affiliated with Oklahoma State University. This organization exists to link advancements in research to the daily lives of rural people through education. This institution has had an immense effect on rural areas and has been instrumental in improving the quality of rural life immeasurably. Historically, however, a strict gender segregation has been practiced. Farmers participated in groups and educational programs geared at agricultural production and career development while county-based women's home economics groups were primarily concerned with providing research-based education for the improvement of the domestic environment. In doing so, this organization promoted a vision of women as primarily consumer specialists. This organization is striving to change its image but these attempts have met with some resistance. Again, Mrs. Becker, "So now they tell us we're educators and not homemakers. And some of us are having a problem dealing with that."

"Why?" I asked.

"Well, my part of it is that I've felt like . . . like I said, I'm a homemaker and that's all I am. As far as I don't have any other . . . well, that's not all I am but . . . that's . . . I'm a homemaker, OK? And that was one organization that I could belong to and feel like I was a part of because I was a homemaker. And now that's changing so much and I don't feel like an educator."

Further contributing to the strong bond between men and farming are women who, also enmeshed in patriarchal discourses of meaning, discredit their own contributions to the farm operation and delegitimize their own labor. Derby (1989) has shown that ranch women in Nevada often do not limit their activities to the familial realm. In fact, she has demonstrated that even though women's work is de-emphasized, their contributions through ranch labor, off-ranch employment, and careful management of family consumption are critical in sustaining the economic viability of operations. Barlett (1993) also found the same to be true in her study of Georgia farm families.

My own study was no exception; in many circumstances women's contributions made the difference between foreclosure and the sustainability of the operation. Many of the women with whom I spoke worked not only in the home, but also on the farm *and* off the farm at a paid job in town. As one of my informants stated, "This makes for a lot of very tired women." Clara McCaffrey was a full-time worker on the farm for many years until financial pressures forced her to seek additional work off the farm. Her list of duties on the farm was impressive and equal to her husband's.

She said, "I've jump-started a tractor with a pair of pliers just like the rest of them. You know, if we were sprigging grass, I was probably driving a sprigger truck or going after spriggs or land pipes, or I'd be driving the pickup if they were throwing pipe off the back. I became a certified AI [artificial insemination] technician and could AI the cattle. If we were gathering and working cattle, I was probably on horseback helping him and of course I helped with shots and worming and castration and any of that that we did. Tried to raise a daughter in those years. I'd get up early, go to bed late, and work hard in between."

I asked her, "Did your husband consider you a farmer?"

"Whadda ya know? I don't know. I never thought about that. I think he considered me a helpmate more than anything."

"Did you consider yourself a farmer?"

"No, I really—I was the wife—and mother."

"But you did all that!"

"Yes, I did."

Contributing to this devaluation of women's work and the promotion of farming as a male occupation was the prevalent notion that farm labor itself was masculine labor. Weston (1998) examined women's experiences entering and working in traditionally male blue-collar trades (such as auto mechanics). Discussing Marx's conception of production, Weston explains the difference between labor—the human qualities managers and coworkers believe will contribute to the accomplishment of job tasks—and labor power, the capacity to perform labor. Logically, the process of assessing an individual's suitability for a job should only involve matching the labor requirements of the job with the labor power of the applicant. Her essential point—a point to which we will return again in this chapter—is that this process is not a simple matter and that assessments of both labor requirements and labor power require that people be seen as possessors of traits of character and competence and that these are infused with cultural notions of gender, race, class, age, and what it means to be able-bodied.

There was a high degree of consensus among my informants that farming was indeed a male occupation. But when we, together, would examine the labor involved, the majority of informants agreed that women could indeed do the vast majority of work. Heavy lifting, of which there is not much, was the exception. As the exchange with Clara McCaffrey indicates, the reality was that a lot of women—both mothers and daughters—were involved in farm work.

But if it is not farm work, what is it about farming that contributes to it being seen as an essentially male occupation? The defining criterion among my informants was management: the person who was making the decisions about the care and maintenance of the land, choice of crops and animals, finances, sowing, harvesting, the sale of crops and animals, and decisions about labor was the farmer. Even though women were involved in many aspects of farming, men were primarily the decision makers in most operations and considered the farmers—and managers. This should not be surprising. As I have argued, men are structurally positioned to be bound to the land as farmers and, as a result, management science education, one of the principal industrializing forces of agriculture promulgated by agents of the state (Fitzgerald 2003), historically has targeted men. This is, however, where the divergence in management practices begins.

The Good Farmer

The postmodern turn in anthropology has meant that recently the discipline has made little use of "depth models" as tools to better understand subjectivity—indicating postmodernism's lack of interest in the innate or essential nature of human subjects. Instead, postmodernism has favored the "contingent" and "emergent" in culture—meaning that history, language, power, and discourses are in constant flux and are mutually dependent; it is difficult or impossible to pin down enduring cultural meaning or truth. This, however, has presented a problem for feminist scholars: for how is it that one can construct a coherent political program of action, or understand change, when one's foundation—what it means to be a woman or a man—is always shifting?

For example, political decision makers often develop policies that reflect the presumption that identities are fixed and immutable, that a unitary or single voice can encompass or represent an entire group of people. Witness the ongoing wrangle over welfare reform. The debate, essentially, is about the personhood of welfare recipients: what they are like (their psyche, motivations, goals, and aspirations, sexuality, desires, etc.) and what they deserve as a consequence. Politics, including those of everyday life, strive to make diversity cogent by creating a "center" of understanding and a stable foundation. This may also be read by many as unitary explanations, stereotypes, incomplete information, or misinformation.

Rather, the answer to the feminist conundrum lies in the politics of everyday life, which becomes central in postmodernist analysis. By "denaturalizing" the workings of power, postmodern feminist scholars resist the attempt to hegemonize identities by recovering diversity and "decentering" the subject—that is, by examining power within the context of particular cultural and historical settings. In doing so, they demonstrate the variety of gendered subjectivities produced: which are valued, which are devalued, and the ways in which different social spheres affect these evaluations. As I have already argued, the subject and subjectivities are understood to be the results of specific historical forms and practices.[1]

The project of decentering the male subject is critical in gaining a better understanding of gender, culture, and power. This assertion is echoed by Weedon (1987:173–174) who argues that:

> The decentering of liberal humanism, with its claims to full subjectivity and knowing rationality, in which *man* is the author of *his* thoughts and speech, is perhaps even more important in the deconstruction of masculinity than it is for women, who have never been fully included by this discourse.

To this end, Rotundo (1993) and Kimmel (1996) have produced histories of American manhood whose principal goal is to question cultural suppositions about gender. Their studies describe the centered subjects that have always been the assumed actors of history and who have always examined, observed, and written about others while themselves remaining beyond the purview of research. In other words, male roles and ideas about manhood, within the context of power and gender inequalities, have been left unexamined until relatively recently. One principal theme that these authors' works share is the significance they give to changing ideas about men through history in any understanding of the current dynamics of American cultural life.

If we want to more completely understand the experience of farm men—their positioning for crisis and the cultural construction of pride—we must examine farm management practices and their implications for gender, identity, and occupational role evaluation in the context of these historical changes. I argue that the appearance and prominence of the highly individualistic values and practices of industrial agriculture are tied to the evolution in American thinking regarding men and manhood—specifically to the emergence of the notion of the "self-made man" that arose, in part, as a consequence of the industrial revolution.

Before 1800, American society, to a large degree, was considered an organic whole, a social order "in which rights and responsibilities were reciprocal and in which terms like individuality or self-reliance had little place" (Rotundo 1993:12–13). Families were regarded as the building blocks of this whole as both the elemental units of production and social reproduction. Men were considered the heads of families and their identities and social standing were closely bound to those of their kin—connected as they were to an unbroken chain of ancestors and descendants.

A man's identity was tied to his performance of social roles and not in self-expression or self-assertion. Duty and responsibility to family and community were the standards of evaluation and men were judged according to their contributions in these realms. There was a strong association between manhood and social usefulness. Men's self-sacrifice was highly valued since it contributed to the maintenance of placid social relations. Rotundo (1993:13–14) states that the ideal man was committed to the betterment of his community: "He performed his duties faithfully, governed his passions rationally, submitted to his fate and to his place in society, and treated his dependents with firm but affectionate wisdom. Pious, dutiful, restrained—such a man seems almost too good to survive on this earth."

Indeed. This notion of an idyllic community prior to 1800, of course, is only partially true. Both Kimmel and Rotundo indicate that self-assertion certainly existed among men (and women). The realities of the daily lives of the new settlers, increasingly interested in western expansion, necessitated some degree of aggressiveness. However, the ideas of individuality and self-assertion coexisted in an uneasy relationship with the prevailing cultural norms of community. But there was flexibility and the possibility of reconciling both the community ideal and individual interests: "Much economic ambition could be rationalized as a man's way of adding to the common wealth, and political self-advancement could always be explained as a desire to serve the community in some greater cause" (Rotundo 1993:15). The salient point, however, is that during this early period of colonial settlement, these individualistic notions were not sanctioned widely and were considered divisive.

During the Great Awakening in the early to mid-1700s, new ideas were reaching the continent from Europe. These ideas advanced the notion of personal independence and challenged prevailing views of social hierarchy. As discontent with British rule increased, these ideas, critical of a static social order and a patriarchal authority, gained adherents quickly. With the Revolutionary War, the concept of self-independence reached a new level of reverence and the essential notion of manhood began to change. No longer was it considered masculine to submit to arbitrary authority. This shift was quite subversive since it undermined the hierarchical social order.

In the late eighteenth century, manhood began to be associated with new traits or "manly passions: assertiveness, ambition, avarice, lust for power" (Rotundo 1993:16). But it also became apparent that these character features could be harnessed to transform society and provide the engine for the development of a new economic system and fresh political ideologies.

The advent of the industrial revolution, from 1800 to 1840, profoundly altered America economically and socially. On the economic front, America's banking system expanded exponentially, a national transportation system began to take form, and new markets, both domestic and foreign, were aggressively developed. A traditional marker of development, participation in non-agricultural labor, rose from 17 to 37 percent of the population between 1800 and 1840 (Kimmel 1996:22). This boom in the economic sector fueled the desires of the nation for western expansion (whose consequence I described in the last chapter), spurred dramatic urban growth and, according to Kimmel, served as a transition point in America's development:

> Such dramatic economic changes were accompanied by political, social and ideological shifts. Historian Nancy Cott notes that the period 1780–1830 witnessed a demographic transition to modern patterns of childbirth and childcare, development of uniform legal codes and procedures, expansion of primary education, the beginning of the democratization of the political process and the invention of a new language of social and political thought . . . Democracy was expanding, and with it, by the end of the first half of the century, America was "converted to acquisitiveness," a conversion that would have dramatic consequences for the meanings of manhood in industrializing America.
>
> (1996:22)

One of the consequences of this "conversion to acquisitiveness" was that notions of individualism, incipient in the eighteenth century, gained currency and power. As mercantile capitalism gripped the nation, one idea in particular held special allure: the equality to succeed or fail. It was thought that competition would reward the best man (Rotundo 1993:19) and, it is important to note, that this evaluative arena was largely male in its composition (Kimmel 1996:26). It was now incumbent upon men to rise and fall according to their own efforts, talents, and attributes. Hard work could take a man as far as he was willing to go and the arena of success was one that, in theory, was open to all men.

This idea resonated particularly well with another emergent belief of the time: that the individual, and not the community, was the fundamental unit of society (Rotundo 1993:19). Men no longer thought of themselves as part of an organic community from which they drew their personal sense of identity. No longer were men linked to a chain of ancestors who would predetermine the boundaries of their influence. Instead, men increasingly identified themselves with their work, a shift that was in line with changing cultural expectations. "The metaphors by which men had defined themselves," Rotundo (1993:19) tells us, "were losing power in the new century."

With the emergence of the notion of the individual, the nascent separation of the private and public spheres of influence began to grow, a change that significantly influenced cultural conceptions of both men and women (i.e., sex roles, male and female relationships, family structure, sex, and social customs) and whose effects is still evident today (see Smith-Rosenberg 1985).

If anything, Faludi (1999) indicates, the social pressure on men for individual success—and consequently familial success—only intensified in the twentieth century. The idea of men as useful to society and the notion of "man in community," she argues, had its last gasps with New Deal America and World War II. In the postwar era, the United States emerged as a global power and began to flex its muscles on the world stage. American political leaders, spurred by the advent of the Cold War, called for America's preeminence is every aspect of life. Men were promised that they would be at the forefront of this movement to demonstrate American dominance. The social pressure was for men to be in control.

Mead, in her study of American society, *And Keep Your Powder Dry*, echoes these themes and illustrates the degree to which these values, spurred by industrialization, had already become institutionalized by the time her book was published in 1942. She explains, for example, that class thinking in America can best be metaphorically described as a ladder: one's position is not fixed for life, but rather one climbs or falls depending on the degree of effort one has exerted. Success in such a situation, she tells us, comes to be defined in terms of how far one has come, how many others one has passed, and what one has in the way of power and possessions. The metaphor of the ladder describes the evaluative standard of success in the farming communities of my study as well.

In northwestern Oklahoma there was not a high degree of class differentiation. In fact, when I asked my informants about their perceived economic class, more than a few were perplexed. Many told me they simply had not thought about that before. "Growing up, we pretty much had what everybody else had," was a common sentiment. It was not that differences were not discerned, but rather the differences that were perceived were of degree and not kind. Instead of social class per se, the dominant, public evaluative basis for success in farming communities was the amount of acreage owned and the public presentation of land and associated properties. Success as a man was defined by his ability to make and enact choices (Errington 1990:639), including economic ones. This proved that he was independent. Farming afforded a lifestyle that supported this independence. Mr. Cobb told me, "You're out in the country, you don't drive in traffic, you don't have neighbors. You can yell and holler all you want to. You come and go as you please, you don't have to punch a clock and you don't have anyone breathing down your neck. You're independent. And when we were making money, farming was a hell of a lot of fun, too."

This independence, however, does not mean that farmers were not conscious and mindful of community standards. Farmers often stated they were perfectionists when it came to their land, that it was important to them how their land looked: that their rows of wheat be even and straight, their fences well maintained, and their watering holes for cattle well constructed. When I asked what an attractive operation conveyed to the community, the typical answer was that it meant that one was a good manager, took pride in one's work, and that, more than likely, one had some extra money in the bank. One's standing in the community as a farmer was based on these evaluations. They were the physical manifestations of good decision making on the part of men. Evaluation was something that was very public; one's land and home were out in the open where everyone could see them, a text that could be easily read by anyone versed in the language of farming. Also, since they were so familiar with the business of agriculture, farmers were able to estimate the financial condition of their neighbors, making one's economic position difficult to conceal.

The Nelsons, for example, bemoaned the fact—especially after they began to have financial trouble—that they lived along a major highway. Mr. Nelson said, "When they [other farmers] would go back and forth between Elsa and Newtown, they were operating my farm, because they would tell me, 'You ought to be doing that' or 'Why are you doing this?' They were telling me what I needed to be doing and so I knew that the community was watching. We all knew that. Because we drive up and down the road and we watch this guy's wheat field and that guy's wheat field and say, 'Oh he's got cheet' or 'He's got rye in his' and all kinds of different things. It gives us a look at what type of farmer he is."

Mrs. Nelson added, "It was the feeling that they would look down their noses at us and that sort of thing, that they would know that we were a disgrace. After all was said and done, we really did have that kind of feeling."

In fact, this mutual observation was highly ritualistic. As I mentioned in the opening, I would ride along with my informants in their pickup trucks and habitually they would slow down—almost to a standstill—and roll down their windows to investigate any changes in a neighbor's land. An evaluative comment by the farmer was always attached to this practice.

But evaluation was not based solely on these criteria; there was room for maneuvering in the public presentation of self and one's farming capabilities. Farmers' stories revealed that individual reputations regarding capability and competence were contingent. As Weston (1998:106) points out in her discussion of women in the trades, "that it is often a symbolically constituted impression of productive capacity that gets the person the job." In Weston's case this often had to do with the job applicant's manner of dress or with behavioral characteristics. Among my informants, stories demonstrating their own farming abilities and large purchases (i.e., evidence of good decision making) were important strategies in helping to enhance or protect one's reputation of competence. No doubt, too, that one's performance of civic and church responsibilities helped to shape the perceived capabilities of men.

Reputation was not only relevant in the public evaluation of status, but also had more concrete implications. Dudley (2000:51–63) argues, like Weston, that determinations made by lending institutions of potential borrowers were not always based on "objective" economic criteria, but were often infused with cultural notions regarding the reputation, personhood, and moral worth of the farmer. The criteria that lending institutions utilized in ascertaining the suitability of loan applicants varied by type of institution. Further, she provocatively asserts that ultimately whether or not a farmer's loan applications would be approved was more dependent on the type of institution from which he borrowed and the evaluative criteria they employed—those based on public assessments of "cultural credit" or individual achievement—than on the amount borrowed (Dudley 2000:63).

Evaluation, however, was not limited to the public sphere; the relative separation of the public and private realms was also the subject of speculation and evaluation in the determination of farmers' "progressiveness" and prestige.

Rosaldo's (1974) noted paper about gendered spaces is important for this discussion. There, she posits a single key explanation for women's lower status in societies. Her study focused on the relative separation of the domestic and public spheres within cultures. The domestic sphere represents woman's domain: the household, reproduction, and the maintenance of family members. The public sphere is the domain of men and includes extra-household labor, citizenship, public culture, and the state. Societies that practiced rigid separations between the private and public spheres—such as Islamic societies that practice seclusion or pre-revolutionary China—would devalue and disempower the private spheres and the women associated with them. Her paper is representative of feminist theory of the 1970s in that it is primarily concerned with accounting for universal gender inequalities and male domination (see also Ortner 1974; Chodorow 1978). In making their cases, these scholars often concentrated on the social construction of the biological "fact" of women's capacity to bear and nurse infants and its psychological consequences.

Rather than assuming that the categories of male and female are endowed with culturally specific characters, feminist scholars today would argue that it is critical to examine "the social and symbolic processes by which human actions within particular social worlds come to have meaning" (Yanagisako and Collier 1987:39). My own research, however, supports Rosaldo's conclusion that the separation of the public and private spheres serves to disempower women and devalue their contributions. Obversely, my research also shows that the fusion of the private and public realms serves to undermine the position of men in the public hierarchy by devaluating their contributions. Both of these assertions are based not on the biological capacity of the sexes, but on the public evaluation criteria employed in rural communities to determine success and prestige—an evaluative mechanism, again, which is enacted primarily in male contexts. For men, the fusion of the public and private spheres compromises the foundation upon which the industrially oriented farming prestige system is based: the capacity of men to make independent management and economic decisions.

A variety of configurations of the private sphere with respect to the public sphere is possible within families engaged in agricultural production. The home is the locus of work for women (although many hold off-farm employment) and, for much of the year, for men as well. Both domestic and farm labor requirements are flexible, enabling involvement of men and women in both spheres and a sharing of responsibilities—at least in theory. Farming, as I experienced it, is seasonally intensive. Ground preparation and the sowing of the crop in the fall and the harvesting of the wheat in late May and early June are the most labor-intensive and time-consuming periods of the year. At these times farmers are rarely at home—day or night. During the interim periods, when maintenance work is not being performed or when cattle are not being worked, quite a bit of business is conducted at home by farmers: making telephone contacts, paying bills, completing other financial paperwork, and coordinating maintenance and farm labor activities. In any event, the work that is required during these periods is not on any strict time frame; farmers can juggle their schedules to suit their needs. Men are present in the home and witness, and sometimes even help with, its upkeep. Even though the vast majority of my informants' children were grown and lived outside of their natal homes, one can imagine that farmers can provide a flexible childcare workforce in the off-season. Barlett (1993) finds this to be especially true among the younger generation of farmers.

The majority of women in my sample worked outside the home, although many did not do so while their children were growing up. When present, women composed a flexible, if exhausted, farm labor force in many cases.

The configuration of the relationship between the public and private spheres varies in its details, significance, and meaning according to farm management style and practices. Though the number of my informants who held more strongly to industrial values was small—too small to draw any firm conclusions—my fieldwork, taken together with the work of other scholars (Collier 1997; Delaney 1991), supports Barlett's (1993) assertion: "that industrialization brings increasing individualism into the expectations about rights and responsibilities between husbands and wives."

My own work and that of other researchers suggest that a relatively strict sexual division of labor was practiced among industrially oriented couples. Men were almost exclusively involved in agricultural production. Women, on the other hand, were oriented away from the operation and seldom contributed farm labor or participated in agricultural decision making. Women were responsible for the domestic sphere. And even if a woman were to work off the farm, as many did, she was much more likely to spend the earned income on herself or on consumer extras.

I worked with Mr. Melrose today. When I arrived at his place, his wife, Nancy, was there. She is an educator in Enid. As I sat down, we began to talk. She is very talkative and was interested in what I was doing.

At one point during our conversation, I asked Mrs. Melrose what type of power women had in the farming operation. She immediately said, "Absolutely none." Mr. Melrose was in the room and he just kind of bristled and mumbled. She continued and said she felt that she had no power because she was completely excluded from any decision making concerning the business.

Fieldnotes, 9 March 1992

Industrially oriented families appreciated the lifestyle and benefits to children that farming and rural life afforded. However, because of the sexual division of labor, women in these operations were much less emotionally invested in, and connected to, the land. This was because farming, to a large extent, only involved them insofar as it provided the resources necessary for the attainment of familial goals and aspirations.

I asked Mr. and Mrs. Becker about women's attachment to land. Mrs. Becker responded that she did not feel that connection, nor did many women of her generation. Mr. Becker suggested that working with the land is partially what creates that connection and that women who are not involved in farm work will most likely not feel it. He said of his wife that "in her married life [she] has not gone out to the barn like her mother did and milked the cows. She didn't have the chickens to take care of. She didn't slop the hogs. She had a garden. You know, that's very important to her, but she didn't drive a tractor." He offered, in addition, that the nature of the attachment to land of more industrially oriented men may also diverge from his own: "I think a lot of the younger farmers and probably the more ambitious farmers or whatever the ambitions are, have become much more businessmen-like and they do not have the relationship with the land that their fathers had and their grandfathers had. Probably they don't care for the land, they haven't taken care of the land to the extent that their forefathers did . . . because of the fact that all they thought about was making the payments on the land that they had and then buying more land."

In fact, unlike their more family-centered counterparts, who see farming as a way of life, industrially oriented women were more likely to view farming as an occupation like any other—a perspective that was not entirely shared by their husbands. For example, one day Mrs. Melrose told me she didn't understand why everyone was making such a big fuss about farmers going under. She felt that they were not very different from men in other industries that were also undergoing economic transitions.

For family-oriented farmers,[2] in contrast, the public and private spheres were seen as being critically linked and interdependent; their synergistic functioning important to farm and familial success. There was indeed a gendered division of labor, but it was more fluid: women performed domestic labor and childcare but, in addition, were more likely to be involved in agricultural work and decision making.

Many women, for example, had special insight into the workings of farming operations because they were responsible for keeping track of financial transactions and thus were in a position to offer management advice. Some women, too, commented that they were the first ones to realize that the operation was heading for financial insolvency because the numbers were simply "not adding up."

Others were integrally involved in farm management. Mrs. Ross, whom I have mentioned before, was the engine that powered the family's farm operation. She was on a constant search for new and innovative strategies to improve production and increase profits. She read extensively and was behind the operation's current experimentation with organic farming. Ironically, on the basis of the hegemonic evaluative criteria utilized by the rural communities of this study, she would be considered a "progressive" farmer. Mrs. Ross's high level of participation in management led Mr. Melrose, her neighbor, to ponder out loud one day who "wore the pants" in that operation. Because of their greater involvement in agricultural labor and decision making, Mrs. Ross and other women in family-oriented operations, expressed a greater attachment to land.

The organizational structure of the American Agriculture Movement, which worked for justice and equity in farm policy and promoted the interests of family farmers, also reflected women's significant level of participation. For example, literally and figuratively, both men and women were at the table, involved in the discussions and work of the organization. Both could vote and women held offices. Mrs. Perkins, who staffed the state crisis line was in many ways seen as the AAM's spiritual leader; she provided the stories of crisis, foreclosures, and suicides that gave the group its reason for existence. More than that, she was a natural leader whose quiet strength commanded respect from all.

Family-oriented women, if they worked outside the home, were much more likely to pool their resources with those of their husbands and have a say about expenditures. Mr. and Mrs. Cobb, for instance, explained how they used Mrs. Cobb's social worker income.

Mr. Cobb said, "What we do—whenever she gets a check, we just gather up all the bills, shuffle them, and start paying them until we run out of money."

Mrs. Cobb agreed, "Yeah, whether it's fuel bills for the farm, repair bills for the farm, whatever, it all has to be paid."

"It don't matter you know, it all has to be paid and whether it comes out of her salary, you know, it's still a dollar in the checking account," Mr. Cobb added.

The language family-oriented couples employed to describe their marital arrangement diverged from that of more industrially oriented couples. On family-focused farms, women were seen as indispensable to the operation. Their participation in farm work, in bookkeeping, their contributions of income through off-farm employment, and their reproductive capacity and maternal roles were instrumental in helping families realize their goal of maintaining the integrity of their operations to pass on to the next generation. As a result, women were seen to *some* extent as equal partners and they were integrally involved in the organization of both family and farm life.[3]

I asked Mr. and Mrs. Gains about the power of women on the farm and in the home. Mrs. Gains said, "I have just as much power as he has."

I asked her why she thought that.

She responded, "It's even-steven. Always has been, I guess. I don't know. I never really thought about it . . . it's a joint effort."

Mr. Gains added, "Yeah . . . it's a joint effort . . . the whole thing is. In our case, why, she's left the decision making and what have you, to me as far as the production system goes . . . 'cause I felt, not too many years ago, the load she had as a mother keeping up with four sons . . . she had a full-time job." Mr. Gains went on to say that the role of women varied from farm to farm and that some women were much more involved in farming.

It was this variation in women's roles that was the subject of some speculation. Since women's involvement was seen as compromising a man's ability to independently make and enact choices according to the hegemonic evaluative criteria, family-oriented men were placed at a disadvantage in the public determination of prestige.

When it came to occupational role evaluations, I could never get a firm response from my informants about whether men who held family-centered farming values—which were held in lower esteem—were also considered less masculine. When I would ask farmers about this, they would always stumble in their responses but would eventually tell me that they did not feel that that was the case. Rather, at work I believe is the mutual structuring of seemingly unrelated arenas of thought that rely on each other for coherence and meaning (Yanagisako and Delaney 1995): the domain of farm management orientations or styles on the one hand, and broader, cultural notions of gender and subjectivity on the other. In other words, it was not that men with particular farming styles were perceived as more or less masculine, but that the values embedded in the industrial management style were supported by beliefs and practices whose development were founded and predicated on a marked separation between the public and private spheres of influence and, thus, gender roles.

Even though farming communities were built on a firm foundation of patriarchy, the erosion, or rather, the shifting, of this foundation by the increasing hegemony of industrial values has not resulted in a narrowing of the gender gap, but in its widening.

In *The Seed and the Soil*, Delaney described the Turkish village in which she conducted research as it made the transition to agricultural production for the market economy. Among other themes, she examined the consequences of agricultural industrialization and its effects on rural life and gender. She could have just as easily been writing about northwestern Oklahoma as rural Turkey:

> As men's relation to land changes, so too does their relation to women. No longer is land imagined as the source of sustenance, a power that demands respect and needs to be placated; now it is seen as a resource to be controlled and exploited. The balance and harmony are destroyed. Women's symbolic value becomes even more secondary and inferior. As men can buy more and more of the items necessary for the provision and maintenance of the household, women are more and more confined to the home, no longer co-(re)producers but consumers.
>
> (Delaney 1991:267)

Even though the industrially oriented women in my sample were not confined to their homes and had greater economic opportunities, their income, as I mentioned, was used for domestic "extras" or for their own consumption. Thus the symbolic and practical result was the same: women had almost no role in the organization of farm life.

Men with an industrial orientation, however, were favored (if they were successful) in the hegemonic evaluative criteria because a man's success was just that: his. When farmers with a more industrial orientation spoke of those with a more family-focused orientation, the former often mentioned the latter's lack of independence to make decisions. Family-focused farmers' emphasis on the values of family continuity on the land and land as a generational trust, and not on profit, was considered noble yet naïve and anachronistic—echoing Mitchell's assertion, alluded to in the introduction, that social beliefs and practices that are incompatible with modernization and its increasing industrialization are "placed in a position outside the unfolding of history" (Mitchell 2000:xiii). In the same breath, these farmers would be derided for their lack of backbone because they did not risk more to expand their operations. It was clear to me that the agrarian goal of family continuity on the land and the strategies employed to achieve it—such as the fusion of the private and public spheres, shared decision making, and the nurturance of the land—were being increasingly marginalized as industrial discourses gained prominence in the region. I argue that this marginalization is itself a hegemonic process by which industrial practices and subjectivities are made to appear independent of family and kinship processes (Yanagisako 2002:13). Farmers who did not subscribe to these more "modern" notions were deemed not progressive or behind the times.

Gender, Pride, and Evaluation

The two management styles discussed here—industrial and family—help us to understand the ideal types, polarities, and discourses that are in competition for authority. What they gloss over, however, is the complexity of daily life. As I have mentioned before, American agriculture is in a period of transition. As a result, most of my informants did not fall easily into either farming model. In fact, of the thirteen couples I interviewed, I would categorize only four as holding strongly to a family-focused orientation (they were all members of the American Agriculture Movement—a family farm activist organization) and only two as leaning heavily toward an industrial orientation. For most couples and other individuals with whom I spoke, aspects of both orientations were evident in their thinking. Indeed, it seemed that it was this coexistence that was so generative of tension during periods of financial crisis as men and women struggled to find meaning and answers and attempted to make sense of a way of life whose fundamental principles and rules they felt had changed.

I would argue that the pride of men is partly tied to the public evaluative criteria of individualism that have come to prominence with industrial agriculture—and to the particular positioning of farmers at this historic moment regarding American notions about manhood: that men, in the equal opportunity world of success, are free to succeed or fail based on their abilities and determination. As families struggled with severe financial pressure, the farmer was labeled as a "bad manager." The label was a mark of individual failure, attributable to his poor management decisions and not to any external factors such as weather, government policies, the economic climate of the nation, the price of the particular commodity, or many others that may influence a farmer's success.

I believe that traditional and patriarchal cultural discourses of rural areas, however, play a much larger role in understanding pride and the responses of men to crisis. A farmer faced with the potential loss of his land found himself positioned in a matrix of meanings regarding failure that implicated notions of land, kinship, history, responsibilities to family (living and deceased) and, as I will show in the next chapter, Christianity. With the emergence of industrial agriculture, these cultural discourses are no longer valued. As the foundation of these traditional beliefs and practices of patriarchy are transformed by industrialization, men who have based their subjectivity and pride upon these agrarian discourses experience their own cultural and emotional devaluation and subjugation. As a result, some farmers withdraw. Some may turn to violence, either self-directed or directed at others. Often there is silence.

Other farmers, however, manage to speak out and tell their stories of the corporate and government institutions they believe have oppressed them and undermined their way of life. Notions about agriculture, rural life and, fundamentally, about the definition of "farmer" have vast social implications. The following chapter profiles an activist political movement that is fighting a battle for the future of rural areas, farming, and the personhood of the "farmer."

The Lord God took the man and put him in the
garden of Eden to till it and care for it.

—Genesis 2:15–16

What we lack, O God, is not of Thy doing but of
man's. In man's greed for gold, he has destroyed
the fruitfulness of the earth. In his lust for power
and dominion he has brought misery upon us all.
The land cries out against those who waste it. Thy
children cry out against those who abuse and
oppress them.

—*Ceremony of the Land*, Reverend Howard Kester and
Evelyn Smith (Munro)

Burn down your cities and leave our farms and
your cities will spring up again as if by magic, but
destroy our farms and the grass will grow in the
streets of every city in the country.

—William Jennings Bryan

FIVE

The American Agriculture Movement and the Call to Farm

I FIRST MET Ann Ross through my work with Rural Health Projects when we were both serving on a committee on rural youth development. During the course of our work and over a period of several months, we developed a warm friendship. One day I happened to mention to her that I had started work on a research project concerning the farm crisis in northwestern Oklahoma. At that time I was not aware of her involvement in the American Agriculture Movement, nor had I even heard of the organization. I consider this historical accident of our conversation fortuitous; not only was I was able to reap the benefits of snowball sampling, but also my relationship with Mrs. Ross provided an immediate and legitimizing entrée to the American Agriculture Movement and its key members.

The American Agriculture Movement (AAM) perhaps most clearly manifests the values of family-oriented farming. This national organization bills itself as the voice of the family farmer, and explicitly positions itself against industrial interests. Mrs. Ross and her contacts were activists and made up the core membership of the Oklahoma chapter.

In this chapter I address the discursive matrix within which agricultural relationships are embedded by exploring the resistance of the Oklahoma Chapter of the American Agriculture Movement (AAM) to prevailing values and structured relationships within agriculture. The beliefs and actions of its members, I argue, challenge hegemonic notions of farming and subvert industrial assumptions about gender, subjectivity, emotions, and the hierarchy of status.

Mrs. Ross introduced me to Alan Gains, the president of the Oklahoma Chapter of the AAM. Mr. Gains was one of the most intriguing people I encountered during the course of my research. We met often, together with his wife, and our conversations were long, a bit convoluted, but always a pleasure. Between our prearranged interview sessions, we would visit informally during meetings or social gatherings of the AAM, to which Mr. Gains always made a point of inviting me. What I sometimes perceived to be our unwieldy conversations I believe now were deliberate attempts by Mr. Gains to highlight for me the interconnectedness of agriculture to every aspect of rural living and, indeed, the broader world. Interspersed with analysis and other voices of the AAM, what follows are extracts from our recorded interview sessions.

I began our series of formal interview sessions with the couple by asking Mr. Gains how long he had been involved in the American Agriculture Movement.

"Since it started . . . I've been involved actually since the twenty-first of December 1977. We done a lot and thought, you know, we knew something was wrong, but we didn't know really what. We thought agriculture was the only thing oppressed, but we didn't get very far down the road until we knew that wasn't entirely the case. We wasn't in the boat alone. And then . . ."

"Who else was in the boat with you?" I asked.

"Well, you know, everybody. We could see early on that laboring people, little business people, middle class people as a whole was really being oppressed with this situation. . . . you know they accused us of being overproductive, and after we got to looking at the records in Washington, D.C. . . . it wasn't overproduction, but it was underconsumption . . . And the records in Washington will show you clearly that never in the history of those records did we have overproduction when we had a strong healthy economy and everybody was gainfully employed, [we] always consumed the product. It's those times when we've got a tilted situation and some major segment of the economy wasn't getting their fair share that the production started piling up . . . So we started trying to find out why . . . Then we decided, hey, those [other] things would be secondary if we just had a balanced economy, if we had parity.[1] We still talk about parity. You know, parity is equality. And they essentially nailed our foot to the floor a long time ago [because] our prices couldn't rise with the cost of living and the cost of production. The products and goods and services that we have to buy, they've increased from 500 to 1500 percent and we're still selling our product at the same old price we had in 1948."

"Now, why is that?" I asked.

"Government policy. Cheap food policy."

"So what you're saying is that they have picked certain industries to target, to keep the price down?"

Mr. Gains responded, "Right, cheap food. People will have more funds to spend on entertainment, luxuries, and motorboats and airplanes and what have you. So that if you spend 12 percent of your disposable income for food—the cheapest as anywhere on this earth—you will advance far beyond anybody else on this earth. And [you] expect the biggest industry on the face of this earth to absorb all of that cheap food? And we just can't do it."

The crux of this conversation is that, according to Mr. Gains, the current poor economic situation of family farmers and their consequent displacement from the land are not the result of the "natural" laws of economics—particularly supply and demand—but considered government policy and, I will presently show, corporate advocacy. To undergird his point, Mr. Gains directed to me to a small booklet distributed by the national AAM office that presents its justification for this claim. The booklet, *The Loss of Our Family Farms: Inevitable Results or Conscious Policies?* (Ritchie 1979), examines various policy proposals from corporate planning groups, principally the Committee for Economic Development (CED), an organization established in 1942 by the presidents of several large corporations and economists from the University of Chicago. This document is worth considering at some length since it provides the discursive framework many members of the AAM used to understand their own economic situation and current cultural predicament.

According to the booklet, the CED is a nonprofit, nonpartisan, and nonpolitical organization. It unites leaders from both business and academia to consider and analyze social issues and research that can serve as guides to the development of public policy. The CED is also concerned with facilitating national dialogues about these matters and promoting better understanding of the American economic system, both at home and abroad. The organization provides a forum that enables business people to "demonstrate constructively their concern for the general welfare" and, in doing so, to help business "earn and maintain the national and community respect essential to the successful functioning of the free enterprise capitalist system" (Ritchie 1979:3).

The CED report of 1962 begins by outlining its view of the problem within American agriculture:

> The common characteristic shared by these [agricultural] problems is that, as a result of changes in the economy, the labor and capital employed in the industry cannot all continue to earn, by producing goods for sale in a free market, as much as they formerly earned, or as much as they could earn if employed in some other use; that is—the industry is using too many resources!
>
> (Ritchie 1979:5)

Once the CED presents its view of the problem—using too many resources—it subsequently reveals its understanding of the cause:

> The movement of people from agriculture has not been fast enough to take full advantage of the opportunities that improving farm technologies, thus increasing capital, create.
>
> (Ritchie 1979:5)

The logic here is that technology provides the capability to perform work with increased efficiency. This statement, I assume, refers to the economic principle of "economies of scale." For example, if there are too many farmers, farm size will be too small to take full advantage of the efficiency offered by agricultural technology. One particular informant told me that working toward achieving peak efficiency was one factor that prevented him from loaning his equipment to other farmers. Any time spent not using his equipment was time spent not taking advantage of its full potential—and thus not maximizing his profits.

Having stated the problem as well as the cause, the CED now suggests the solution—an "adaptive approach":

> The adaptive approach utilizes positive government action to facilitate and pro-
> mote the movement of labor and capital where they will be most productive and
> will earn the most income. Essentially this approach seeks to achieve what the
> laissez-faire approach would ordinarily expect to achieve but to do it more
> quickly and with less deep and protracted loss of income to the persons involved
> that might result if no assistance were given.
>
> (Ritchie 1979:5)

This approach necessitates programs and policies that will attract resources from the agricultural sector of the economy and mitigate the effects of the adjustment on people and property:

> If the farm labor force were to be, five years hence, no more than two thirds as
> large as its present size of approximately 5.5 million, the program would involve
> moving off the farm about two million of the present farm labor force, plus a
> number equal to a large part of the new entrants who would otherwise join the
> farm labor force in the five years.
>
> (Ritchie 1979:6)

But one of the main obstacles to the realization of this goal, according the
CED position, is that the "support of prices has deterred the movement
out of agriculture" (Ritchie 1979:6). To address this problem, the CED
recommends lowering agricultural prices:

> The basic adjustment required to solve the farm problem, adjustment of re-
> sources used to produce farm goods cannot be expected to take place unless the
> price system is permitted to signal farmers.
>
> (Ritchie 1979:7)

Further, they recommend that:

> [T]he price supports for wheat, cotton, rice, feed grains and related crops now
> under price supports be reduced immediately.
>
> The importance of such price adjustments should not be underestimated.
> The lower price levels would discourage further commitments of new produc-
> tive resources to those crops unless it appeared profitable at lower prices.
>
> (Ritchie 1979:7)

According to the CED, this movement of resources out of the agricultural sector would increase sales of these products at home and abroad due to their decreased costs. Cheaper commodities would be possible because of their disentanglement from the web of trade incentives, disincentives, and other "inconsistent policies that now surround our foreign trade in these farm products" (Ritchie 1979:7).

Based on the argument thus far presented, Ritchie (1979:8) draws a number of conclusions. He reasons that such a policy will result in: (1) increased returns on corporate investments in agriculture, (2) over two million farmers and families entering the urban labor pool, which would tend to depress wages, and (3) lower prices of agricultural products, which would both increase foreign trade and provide cheaper raw materials for domestic food and fiber processors.

Before I had begun to better understand the dynamics of agricultural modernization, I found Ritchie's conclusions surprising. My initial impression of the report was that the CED was presenting a rather disinterested assessment of economic policy on a national scope. The consequences of their recommendations are considerate of economic imperatives but not necessarily of the human reality or the personal consequences of such policies. And this last point is at the crux of Ritchie's argument and conclusions. His conclusions, in contrast, are decidedly interested and reflect his concern for the maintenance of farm families on their land and the agrarian way of life.

Later in the booklet, Ritchie lists the influential members of the CED and their affiliations and provides a short descriptive narrative of each to show their connection to government and agricultural policy development. He does this to demonstrate that the membership of the CED may not be entirely disinterested either. For instance, many representatives of food corporations hold memberships in the organization. In addition, Ritchie shows that several members of the CED and its staff and research team were the "very ones making government policy at the time, and who would implement the CED's recommendations during the next 15 years" (Ritchie 1979:10).

One metaphor, used universally by my informants who were members of the American Agriculture Movement, is worth quoting here. This metaphor was often employed by the subjects of this research to describe the misdeeds they believed the government, at the behest of corporations, was perpetrating on family farmers. The example I take is that of Kenneth Boulding of the University of Michigan Department of Economics. This member of the CED is described as an "influential agricultural economist, who once wrote":

> The only way I know to get toothpaste out of a tube is to squeeze, and the only way to get people out of agriculture is likewise to squeeze agriculture. If the toothpaste is thin, you don't have to squeeze very hard, on the other hand, if the toothpaste is thick, you have to put real pressure on it. If you can't get people out of agriculture easily, you are going to have to do farmers severe injustice in order to solve the problem of allocation.
>
> (Ritchie 1979:12)

A 1974 report of the CED, assessing the progress of the organization's 1962 plan, is also highlighted in the booklet. In essence, the report reveals that their plan had been successful:

> The situation of U.S. agriculture has changed drastically within the last decade. In 1962, when the Committee issued the policy statement *An Adaptive Program for Agriculture*, the problems of U.S. farming were mainly related to maintaining farm income in the face of continuing surpluses. The diagnosis was that agriculture was using too many resources; fewer farms and farmers could produce all the output then required or even more than could be marketed. As a result of these findings, we prescribed programs for the better use of our resources in agriculture that, vigorously prosecuted, would enable the people involved in farming to receive higher incomes without government controls or subsidy. In general, policies of this nature have been pursued by the U.S. government, with the result described in the present statement: namely, that U.S. agriculture today is a far more efficient, far more productive industry.
>
> (Ritchie 1979:13)

The report declares that the government instituted their policy recommendations, that their policy was correct, and that their goals were achieved. They continue:

> The farm population is now so small in relation to the total population that further migration from farms will not be substantial. Annual agricultural employment, which was 4.5 million persons only 10 years ago, is now about 3.5 million persons, or only 4 percent of the total labor force, and it is still declining. It represents approximately the optimum labor force that this Committee envisaged for the 1970s in its statement *An Adaptive Program for Agriculture* (1962).
>
> (Ritchie 1979:13)

According to the 1974 CED report, the result of these policies has been the creation of a two-tiered system of American agriculture:

> One sector consists of large farms that, although numbering about 23% of all farms, produce 80% of all farm marketings. This group is engaged in the production of the major food and fiber crops such as grains, oilseeds, and cotton. These products of commercial agriculture have been the focus of U.S. agricultural policy over the past 40 years and are now the major crops in world trade.
>
> The other 75% percent of U.S. farms, accounting for only 20% of the output, are operated largely by farmers who are increasingly dependent on the industrial and service sectors of the economy to provide supplemental or full-time employment. Where financial distress exists in this group, it is rooted mainly in general social and economic causes, not in farm prices. Assistance for these farmers should be extended not through special support programs but rather through the same kind of program that should be made available to all disadvantaged Americans, urban or rural, i.e., through a national welfare assistance program based on a minimum annual income.
>
> (Ritchie 1979:15)

This is significant, according to Ritchie, because the policies pursued by the government have served to support the goals of industrial agriculture, while undermining the work of family farmers through the enforcement of "below parity prices." Their policies have encouraged the development of a two-tiered system within agriculture that is plagued with inequities. Because of this, "They [CED] make a strong point that the possibility of 'a recurrence of agricultural instability' must be kept in mind so as to maintain 'an atmosphere relatively free of the political pressures from farmers experienced in the past'" (Ritchie 1979:18).

For Ritchie, this is a critical point. He sees the political strength of family farmers as the real target of CED recommendations and government policy. He argues that curbing the power of farmers is significant for corporations for two related reasons. First, farmers historically have aligned themselves with trade unions and workers, forming farmer-labor alliances that have been effective in restricting the control and power of corporations. Second, farmers have been at least partially successful in their ability to advocate for agricultural policy that has been favorable to agrarian interests (Ritchie 1979:18).

If corporations and the government have conspired to constrain family farmers' power, the birth of the American Agriculture Movement has been the response. The movement, in essence, was formed to bolster the political strength of farmers and address the disparity in the structure of agriculture.

The term "structure" refers to the entire range of factors that serve to pattern relationships within a particular industry. In agriculture it may refer to such issues as ownership patterns, financial arrangements, the role of workers, access to inputs or resources, competitive relationships, and so forth (Strange 1988:56). The structure of agriculture has been a topic of tendentious debate and controversy since the 1930s, a time when concern with increased farm tenancy and the unbalanced distribution of wealth led President Roosevelt to form a special committee to address these issues. The committee's work ultimately led to the development of New Deal programs to redistribute land and support owner-operators (Strange 1988:57).[2]

In the 1940s, anthropologist Walter Goldschmidt added to the growing debate about agriculture and its structure. His classic study, *As You Sow*, compared two communities of similar size: one had an agricultural system composed largely of midsize owner-operators and the other of large absentee-owned farms. His findings were stark in their contrast: (1) The small farm community supported nearly twice as many businesses as the industrial-farm community and two-thirds more retail trade; (2) spending for household supplies and building equipment was three times greater in the small-farm community; (3) there were 20 percent more people per dollar of agricultural crop sales in the small-farm community, where half the breadwinners were self-employed, whereas two-thirds of the people in the industrial-farm community were agricultural wage laborers, with fewer than one-fifth of paid workers being self-employed; (4) the small-farm community had four elementary schools and a high school while the industrial-farm community had only one elementary school; (5) the small-farm community had three parks and two newspapers compared with one corporate-owned playground and one newspaper in the industrial farm community; and (6) the small-farm community had twice the number of civic organizations, churches, and churchgoers (Strange 1988:86–87).

Goldschmidt's findings were controversial since they did not paint a kindly picture of the industrial farming enterprise, which had long been entrenched in California (Wells 1996). As a result he was asked to leave the Bureau of Agricultural Economics (who had sponsored his research) and his findings were politically suppressed (Goldschmidt 1978).

In the 1960s, a group of agricultural economists from various land-grant universities began to examine the structure of agriculture and, specifically, the emerging phenomenon of industrial agriculture. The principal question they addressed in their inaugural booklet, leaflets, and essays was, "Who will control U.S. agriculture?" In various other publications they presented a series of scenarios providing diverse answers to this question. The debate on structure, however, remained a largely academic one.

It took the formation of the American Agriculture Movement and their public protests in 1977 through 1979 to galvanize public interest in the issue of structure (Strange 1988:58). According to their own history (AAM, n.d.), the American Agriculture Movement was born "out of desperation" in the fall of 1977 after Congress enacted a farm bill that "insured four more years of prices paid to farmers below their cost of production."

The initial protest took the form of a national farm strike. AAM members threatened to cancel their memberships in all farm organizations that did not support them and, more importantly, pledged not to plant the 1978 crop unless Congress enacted legislation that would guarantee farmers parity. Their demands were not met and their strike gave way to more general protests, including the famed "tractorcades" and an angry protest in Washington, D.C. in February 1979.

Many of my informants were involved in these protests at the local and national levels. They recall this period with a degree of nostalgia, but only because support for changes within the agricultural sector seemed to be more broadly based and participation in demonstrations—and in the AAM—greater.

Mr. and Mrs. Cobb for instance, had been involved with the AAM early on, but no longer were. I asked them if there had been a lot of farm activism in their area in the late 1970s.

Mr. Cobb responded, "Yeah, there was. There was a lot."

"We were in Oklahoma City with a tractor one weekend," added Mrs. Cobb, referring to a tractorcade in which they had participated.

Mr. Cobb continued, "Yeah, Senator David Boren was governor at that time and he came out. It was colder than heck that one day and [he] stood out on the street and waved at every tractor that went by. It was great . . . I remember that. But there was a lot of farmers around here that took part in that that had never done anything before. It was a group [the AAM] that came along with an idea that fit for the time, place, and people. It was something for a lot of people that didn't fit any other farm group."

Mrs. Cobb added, "They didn't. They always stayed on the farm, hadn't been involved in anything outside of their acreage. You know, it got them out."

I asked them, "What got them out? I mean . . . you were saying before, the lower prices?"

Mr. Cobb answered, "Yeah, low prices and grain embargoes and stuff like that, it just . . . as they say on the farm, it finally pissed them off."

"The people working out there finally realized they weren't getting paid for what they were doing and they still aren't . . . fewer of them [now], larger farms now," Mrs. Cobb said.

The tractorcades are a dim memory from my early high school years. Typically, farmers, in their tractors, would embark upon a slow journey to a central location, such as a county seat. Along the way, other farmers and their machinery would join the convoy. One distinct memory that I do have is trying to make my way to the town square of Enid (then the commercial center of our town) only to have my way blocked by dozens of tractors. The sight was impressive. Demonstrations like this one were intended to raise awareness of the predicament of farm families, but also to illustrate the connections between the farming and town economies.

Nationally, Mr. Gains was involved in a tractorcade that descended upon Washington in February 1979. The story of this tractorcade, included in the AAM's own history (n.d.), has taken on a somewhat mythic air:

> Farmers driving tractors left the countryside, collected along the highways, and traveled in convoy to the nation's capitol at 15 miles per hour. The trip took 18 days for some as the tractors could only travel about 100 miles per day.
>
> Through snow, sleet and mud they pressed on, camping along the road when necessary, stopping in parks, shopping center parking lots or a farmer's field when available. They endured a blizzard in the mountains, were snowbound in Illinois and inched their way through city traffic in large cities like Dallas and Atlanta. For the farmers, it was a harrowing but memorable event.

Mr. Gains picks the story up from here: "The people were sympathetic after they realized what our efforts amounted to. And all the way up there, you know, they had a guy went ahead, you know, and made the arrangements where they were going to stop. And the people would have food and a place to park and everybody sympathized with us. And it was quite a sight, if you can imagine 2,000 tractors in one bunch and they all had the American flag and their state flags and the American Ag flag."

"We have a lot of pictures," Mrs. Gains said.

Mr. Gains added, "And we have a tractor today in the Smithsonian Institution. They came to us and wanted one of those tractorcade tractors."

From the AAM history:

Parade permits had been secured once they reached Washington. The citizens and the police force could not handle it, however, when thousands of farmers on tractors converged on the city from four directions on the morning of February 5, 1979.

Mr. Gains continued, "And we went down in front of the capitol, at the reflecting place and parked on the north side; you've been there. That was the farmer's market. Everybody had a sack of corn or wheat or oats or something in their tractors and they was going to pull up there and this would make it all legal. We had all the permits, everybody's blessings, and doing it that way was like he was taking his stuff to market. Andrew was there [a key member of a radical offshoot of the national AAM, about whom Mr. Gains had spoken to me previously]. Had a bright red pair of insulated coveralls on, and his face was as red as his coveralls. And when the tractors came around that corner and started to make the turn back around the reflecting pool, he stopped them. Got out there in the middle of the street. If I'd been big enough to have done it, I'd probably tried to carry him out of there. But that's when the fracas started. It wasn't, you know, the policemen up there. You see them real readily. But I mean to tell you, when he done that they was policemen come out of everywhere. They must have been in the trees, in the buildings, in the . . . out of the grass and that was awful."

"What happened?" I asked.

"Sick. Oh, got an awful fracas. You know, they . . ."

"Tear gas," Mrs. Gains jumped in.

"Yeah, a cop went up to that one tractor and shot some sort of tear gas projectile through the glass. He had the doors locked and then, when they done that . . ."

"It blinded him."

Mr. Gains continued, "He beat it out and threw one, about a half pound can of tear gas in that tractor with him. Just filled it full . . ."

"See, that's things you didn't hear about . . . he lost one eye," added Mrs. Gains.

"He lost one eye, and then they got him out of there. We all, you know, I just, I couldn't believe my eyes. They drug him out of there and put him in the back of a pickup. One of them goons hit him on the hand. Had his hand down on the bottom trying to hold himself. And they struck his hand. He lost one eye and, of course, that was just one of several incidents where people got hurt."

Mrs. Gains said, "We had a friend [in another town] that lost a kidney over it. He'd been . . . ruptured his kidney."

"Yeah, jabbed him in the back with a night stick and it ruined a kidney."

Referring again to Andrew—the individual who began the disturbance, and who we will meet again in this chapter—Mr. Gains continued, "But anyway, that's the frustration afterward that he felt. The president had denied him, which was Carter. He was supposed to do so much and he just turned his back on him and left him standing. Why then, I was sympathetic for him, but I didn't want anything like that to happen, you know, it wasn't planned. But that's how determined he was. He said, 'They've inconvenienced us for years and it ain't going to hurt to inconvenience those bastards,' that's the way he put it, 'for just a few hours.' And it did. But it caused an awful traffic jam, my word . . ."

AAM history:

> Before the day was out, 19 farmers had been arrested, 17 tractors impounded, and police had penned the farmers into an involuntary prison camp on the Washington Mall.

The Carter administration was indeed the focus of farmers' and, specifically, the AAM's, ire. The organization's relationship with the administration's secretary of agriculture was particularly contentious. Bob Bergland's attitude toward agriculture could be characterized as laissez-faire: he perceived government policy as having little effect on the structure of agriculture, reasoning that economic market forces would eventually work out any problems that farmers were experiencing. Especially inflammatory was his characterization of the AAM membership as "high rollers," individuals who had heavily indebted themselves to share in the bounty of the expanding agricultural economy—but who misjudged and were now paying the consequences of their error (Strange 1988:59).

Two reports, one by Congress's General Accounting Office and the other by the Congressional Budget Office, contributed to Bergland's conversion regarding the role of government in agriculture's economic structure. Both reports signaled that government policy may indeed be responsible for the increasing and undesirable trend toward the concentration of agricultural production on fewer and fewer farms (Strange 1988:59). Because of the reports, Bergland organized a detailed and thoughtful study of agricultural structure that included public hearings with, and testimony from, farmers around the country. This would be Bergland's legacy: "a clarion call for reform of agricultural policies that produce unintended consequences in the economic structure of agriculture" (Strange 1988:60).[3]

The report that resulted from the study, *A Time to Choose* (U.S. Department of Agriculture 1981), was significant. For one, the report legitimized the question of who benefits from agricultural policy and government farm programs, a question with immense political and policy consequences that had gone largely unexamined.

Second, the authors of the report created a conceptual tool, the "three farms" model, to understand the dynamics of structure within the agricultural sector. The three farms approach categorized farms according to the volume of farm products each sold. This analytic framework, based on the relative size distribution of farms within the United States and its relation to policy, has persisted. The three farms, in gross terms, are as follows (Strange 1988:62–64):

- *Small farms*, according to the conventional thinking advocated by the report, are not really farms at all since most of their income is generated from off-farm employment.
- *Medium, or midsized farms* are chiefly managed by family farmers (here I refer to the unit of production, not the particular family's management orientation). These farmers receive a portion of their income from farming, but usually have to rely on off-farm employment as well. For instance, all of the farm operations of the couples I interviewed for this study fell into this category.
- *Large farms* are fully modern, capitalized, technological, and efficient. Even though they comprise only a small portion of the total farms in the United States, they produce the lion's share of agricultural products. The income generated from these operations place their owners in the high-income brackets.

Finally, the report presented a number of the study's conclusions about the consequences of government policy and their relative effects on the three farms described above. These included (Strange 1988:61):

- Tax policy is biased toward larger farms and wealthy investors.
- The marketing system is increasingly oriented to serve larger producers.
- Commodity price support programs and credit services have benefited larger producers and landlords.

- There is little or no efficiency to be gained from the further expansion of large farms. As farm size continues to grow, additional labor is required, which is very inefficient and increases the per unit cost—resulting in a "diseconomy of scale."

So that I might study the local effects of agricultural policies and restructuring, the Oklahoma Department of Agriculture provided me with data regarding the distribution of farms by size and through time, from 1959 to 1997, for the eight counties in which I conducted research. For the purposes of this exercise, the following acreages were used as criteria: small farms included those between one and 259 acres; midsized farms, between 260 and 999 acres; and finally large farms were those that were composed of 1000 acres or more. Again, my principal interest was in midsized farms,

Distribution of Farms by Size, Eight-County Area, Northwest Oklahoma, 1959–1997

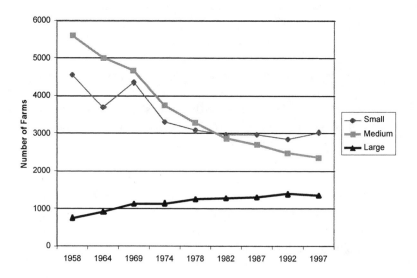

Source: U.S. Census of Agriculture, U.S. Department of Commerce, Bureau of the Census. Obtained from the Oklahoma Department of Agriculture, Food and Forestry, Statistics Division.

the category that contained all the farms of the families taking part in this study.

As the accompanying graphs demonstrate, the effect has been dramatic. In 1959, there were 5,599 midsized farms in the eight northwestern Oklahoma counties of interest, constituting 51 percent of the total number of farms. By 1997, that number had plummeted by 58 percent to 2,349 farms. Midsized operations now made up only 35 percent of total farms. There has also been a substantial drop in the number of small farms, from 4,547 in 1958 to 2,349 in 1997, although the proportion of small farms has remained relatively stable. In 1958, they constituted 42 percent of farms, climbing to 45 percent of farms in 1997. The only substantial increase in the absolute number of operations was in the large farm category. In 1958, there were 763 farms that were over 1000 acres, constituting seven percent of total farms. By 1997, the number had increased by 77 percent to 1,353; large farms now made up 20 percent of farms in the eight-county area.

Proportional Distribution of Farms by Size, Eight-County Area, Northwestern Oklahoma, 1959–1997

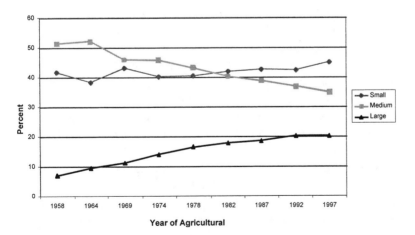

Source: U.S. Census of Agriculture, U.S. Department of Commerce, Bureau of the Census. Obtained from the Oklahoma Department of Agriculture, Food and Forestry, Statistics Division.

From this data and that presented in chapter 3, it is evident that the to-tal number of farms is decreasing, with medium and small farms being the primary casualties. While the proportion of small farms has remained fairly even, the percentage of medium-sized farms has declined significant-ly and the proportion of large farms has increased substantially. In short, midsized operations have been most significantly affected by the policy en-vironment and other changes during the course of the last 45 years.

When agricultural policy is considered from the perspective of the three farms paradigm, it becomes more apparent why some of these changes in the structure of agriculture have taken place. From the policy—and industrial—angle, large farms are relatively easy to deal with. They need little public support as they have captured the benefits of technology and economies of scale. Since they are heavily capitalized and rely on bor-rowed money, they do require policy that encourages a steady flow of ac-cessible cash. They are also subject to the affects of price variation and to fluctuations in demand. As a result, these operators are proponents of gov-ernment policies that bolster the expansion of markets and promote stabil-ity within the agricultural sector. Finally, these operations are sensitive to general economic conditions to a degree that smaller farms may not be (because their operators can rely on other sources for income), and thus these producers are concerned with macroeconomic policy that fosters general economic security and progress.

According to the conventional wisdom, small farms are even easier to deal with. They are not really considered farms at all, since so much of their income comes from off-farm labor. No serious agricultural policy considerations are geared to them, and their needs are best met through the development of social, and not agricultural, policy.

According to institutionalized views, midsized farms, on the other hand, present a dilemma. They are viewed—in psychoanalytic terms—as being in a state of arrested development. If they really are farms, their "natural" course of evolution would be toward achieving economies of scale and increased efficiency, meaning growth: more land and more technology.

There are three schools of thought regarding agricultural policy and family farmers. One view is that of the CED reports: these midsized farms can opt for growth to take advantage of technology and economies of scale; if they do not make that choice, these family farmers should be helped out of agriculture in a way that eases that transition, and that allows for the continued growth and progress of the industry toward technology and efficiency. Many classically trained economists and politicians take this position.

A second viewpoint is espoused by the AAM and other farm activists groups that organized in the 1970s and 1980s. Their primary concern is with saving the family farm. They understand family farming as a tradition that is worth preserving. To that end, they advocate for commodity price supports to guarantee that these operations will be able to survive.

Finally, many liberal economists, politicians from farm states, and some church and rural advocacy groups (including Strange's Nebraska-based Center for Rural Affairs) believe that it is nonsensical to support commodities at the same price for every farmer regardless of the diversity of operations (e.g., farm size and needs). Instead, they advocate for supplements paid directly from the U.S. Treasury and carefully targeted to those operations with income needs.[4]

Strange (1988:75–77) finally posits that the three farms model may be of limited value. He argues that there are methodological problems with the report. But his principal apprehension with *A Time to Choose* is that it is infused with the values of industrial agriculture. Its focus on the problematic of family farming, he argues, tends to divert us from a closer examination of the problems immanent to industrial agriculture. The report's privileging of the volume of farm product sales as the chief variable in the determination of farm structure dismisses many other equally important variables, including power relationships among operators and owners, management structure, the land base, the tenure of the operator, the farm's financial structure, and the diversity of crops raised. Finally, the report assumes that the policy needs of dissimilar farms can be met by dispassionate policy makers. Farm practices, as should be clear by now, do not exist in a power vacuum; cultural, economic, and political interests are tied to, and influence, divergent farming strategies.

Perhaps the greatest achievement of the American Agriculture Movement—through its engagement with government and advocacy for mid-sized family farms—has been to bring the question of structure and relative policy benefits to the foreground.

"So, it's planned," Mr. Gains told me, "and I make a concession to all those people that made those plans and created that policy. Maybe they think of the best interest of the nation as a whole, but I got an argument with it. For young people like you, I think you'll undoubtedly see the day, if it goes corporate, that you may give five dollars for a loaf of bread. 'Cause when the corporations get it and run it like a business, as they run everything, there's going to be a return to every dollar that goes into it, or they're not goin' to do it."

Keeping corporations from "doing it" and maintaining family farmers on the land was the focus of the activities of the Oklahoma Chapter of the American Agriculture Movement.

Yesterday I went to the AAM meeting that was held at K-Bobs (steakhouse) in Wood-ward. I think it's interesting that the "state" meeting is often held in the northwest part of the state. What does this mean? That all the leadership and most activists are here? That's the way it looks to me.

The meeting began with a call to order. There were about 35 people present. People introduced themselves. I thought that it was interesting that only the men stood up and introduced themselves and their wives. People were there from all over northwest Oklahoma. After introductions they had a prayer and the flag salute. Minutes of the last meeting were read (about 34 were at the last meeting).

Before I go on, I forgot to point out that the women of AAM are putting together a quilt. The quilt will have a section from every county in OK that has AAM members. The women wanted to see what other counties had done. In most cases the quilted sections contained the name of the county with an embroidery of something significant about that county. In other cases sometimes a picture of one of the demonstrations was contained—with people from that particular county participating.

After the minutes of the meeting were read they moved right on to old business. They gave a financial report. They have what they call a parity fund. This is a fund to support the national and state offices of the AAM. People put in $1,000 or $500 individually— to draw interest. They currently have $28,000 in a CD. They have $1,700 in a checking account.

They talked briefly about the fish fry that National AAM holds. This is a legislative reception that apparently has grown to be one of the biggest in Washington, D.C. It was a "howling success" and a few Oklahomans went up for it.

Fieldnotes, 7 March 1992

After the tractorcade of February 1979 in Washington, the AAM established a national office. In fact, much of the activity of the AAM seems to have shifted from the tractorcades and demonstrations of the early days to targeted political advocacy in an effort to make their voices heard in the arena of public policy development. Their increasing sophistication in this area is evident. At the time of this research, the AAM employed the standard legislative advocacy strategies to maximize the impact of their message on policy makers. They hired lobbyists to work with members of Congress. The national office established a political action committee (PAC) to monetarily support candidates for elected office that were sympathetic to their cause, including "higher minimum commodity prices, inventory control, and supply management" (Summers 2001:314). They organized an annual fish fry in Washington to host policy makers in an informal setting and to build more personal relationships with officials. The AAM's manifold strategies were directed toward one goal—parity: "We want to get fair prices in the *marketplace*, not through subsidies" (AAM national brochure, n.d.).

Finally and significantly, the AAM delivered the stories of real farmers to policy makers and to the nation at large. When Congress held hearings on agricultural issues or rural life, the AAM ensured that family farmers were represented. They coordinated visits to Congress when AAM members traveled to Washington. Additionally, national and state leaders (and individual members as well) of the organization placed editorials and letters to the editor in newspapers across the country.

At the local level, their success, in my opinion, was not in the field of public relations (some of my informants who were not members of the AAM, but who had been involved at one time—the Nelsons and the Cobbs—believed the movement had all but died out), but rather in helping farmers frame their own experience of financial struggle, bankruptcy, or land loss in a way that was less personally damaging.

I asked Mr. and Mrs. Gains, "What do you think men are thinking about when they're going through the crisis? What makes the burden so heavy? What are their preoccupations?"

Mr. Gains responded, "I think the big thing is that, like I said, they feel like they've failed. They didn't do enough . . ."

"Of what?"

"They didn't produce enough. [They] lose sight of the fact that the reality of that is that they did produce enough, but trying to sell it at the price that they sold it at wouldn't make ends meet. But still they sell themselves short. I know that's been the case in several of them. This old fellow . . . he was real nice, real jovial, easy to get along with, real considerate and I know good and well that's the way he felt. He accepted the fact that they say, 'You've been inefficient and you've been a bad manager.' And he went out there in a pasture and blowed his head off. And that's the sad part, to me, is him dying or anybody like him, is the thought in his mind that he was a total failure. And he had produced train loads of food of every kind."

I asked Mr. Gains, "Do you think men feel blamed?"

"Yeah, well, I think they take the blame on themselves. And at the time, see, this has been about five or six years ago, that was still real common. To hear that was real common. 'Those inefficient farmers. They were inefficient, and they're bad, bad managers.' You heard that from everybody. [The] Farm credit system was big pushing that. And it wasn't their [the farmers] fault. No way, the most inefficient ones there was, it wasn't."

Part of the framework the AAM supplies to help farmers understand their own experience and tell their stories is that losing one's operation is not necessarily one's fault. To them, documents like the CED reports serve as evidence that the movement of farmers off the land was planned or, at the very least, that current policies do not support the goals of family farmers.

The AAM highlights many reasons why a farmer might lose his operation. These could include weather, banking policies and practices, and agricultural policies. The group also highlights why some farmers may be especially successful—including having an oil well on one's property, which a couple of my informants were lucky enough to have. This makes the AAM vision, given the scientism of industrial agriculture, quite subversive. The movement challenges the rural social hierarchy of status by destabilizing the equation of "successful farmer/man = independence = lots of land," by showing that men are not completely independent, and that the success or failure of their operations is as much predicated on factors beyond farmers' control as it is on their prowess in making the right management decisions.

Next Mrs. Perkins gave a short talk about her recent experiences with the hotline. Her talk, it seemed to me, was basically aimed at making us realize that there is still a problem. She pointed out that in mid-March FmHA (Farmers Home Administration) sent out between 1500 and 2000 foreclosure notices—that this didn't include people becoming delinquent as of Jan 1, 1992—these people will get them later. She said the past two weeks the calls have been very bad. She said, "We actually have terrorism going on because when we get one of those letters, terror strikes our hearts."

I'm not sure if she said that they went, or were going to go, see Gary Shearer (Oklahoma State Secretary of Agriculture) and the Governor and plead to continue the Governor's Task Force. During the time when the task force was functioning the number of suicides apparently went down.

Next they announced a farm show 10–12 of April in Oklahoma City at the fairgrounds—I think they were going to have a booth.

Next the Ag in the Classroom program was discussed. The program is K–12 grade. Barns will be provided which will have hands-on ag materials. An objective of the Task Force 2000 was to increase ag literacy. Everyone depends on foods and fibers—people should know something about them. It seems to me that their interest in this program is more evidence that AAM is not just interested in the farming crisis, but on preserving a whole way of life that they feel they are losing. "We have deep convictions—deep as our roots in this old soil," someone said during the course of the meeting.

Fieldnotes, 7 March 1992

Many members of the AAM were closely connected to the work of the AG-LINK hotline. Several volunteered as first responders. They viewed their participation as a natural extension of their work with the AAM— helping to keep families on the farm and using their own experiences to help those in distress to reframe their experience and find some hope.

When a farmer somewhere in the state was perceived as being in imminent danger of perpetrating violence against himself or others, a first responder was dispatched to the farmer's location. Mrs. Perkins relied on her extensive network of professionals and trained volunteers to reach out to the far corners of the state. She also used these contacts to keep her up-to-date with local communities. For example, I attended a Methodist men's gathering where Mrs. Perkins was to be a featured speaker. Over dinner I told her that I had just heard about the suspicious death of a farmer in a particular community. She immediately responded that she had cousins in that town and "I'll see what I can find out."

As I mentioned before, Mrs. Perkins served a critical role in the AAM. First, she provided leadership: she was a quiet powerhouse and people listened to her opinions and looked to her for guidance. She was deeply respected. Second, she and her family had lost their farm. Her husband died of a heart attack in her arms, an experience she attributed to the chronic stress her husband had endured as a result of being under severe financial burden. This gave her substantial credibility within the organization. Finally, as the head of the state agricultural help line, she was at the center of the crisis, probably the single person in the state who had most knowledge of the impact of the financial crisis in the agricultural sector as it was experienced across Oklahoma. Her stories of troubled farm families kept the AAM membership mindful of the continuing problem and the importance of their mission.

In addition to their participation in the work of the AG-LINK office, AAM members had also been involved with the task force on farm-related deaths that had been organized under Governor Henry Bellmon, one of the few true statesmen of Oklahoma. Governor Bellmon was from a farming family in northwestern Oklahoma that had come to the region immediately following the Land Run. He formed the task force in response to the report mentioned in the introduction that indicated that the leading cause of death among Oklahoma farmers—previously thought to be tractor rollovers—was suicide. The task force coordinated a series of ten hearings across the state to learn about farmers' experiences. The transcripts of the hearings were sent to policy makers at the national level, including members of Congress. The task force was also responsible for changing policies that enabled farmers in need to qualify for U.S. Department of Agriculture (USDA) programs, including, ironically, food stamps.

A long-term strategy the AAM pursued to keep family farmers farming was the development of farm literacy programs. During one of our interviews Mr. and Mrs. Gains and I talked about the "Ag in the Classroom" project. At this point in the conversation we were discussing the fluctuating price of wheat and the American public's lack of understanding regarding agricultural dynamics. Mr. Gains said, "The price of wheat went up, you know, people [thought], my goodness, wheat, a loaf of bread's gonna go out of sight. And a lot of people, when they questioned it, didn't even know what in the world was in a loaf of bread. What kind of grain, what raw material it took to make it. And I blame our government for this to a high degree. The possibility of the 'Ag in the Classroom' is tremendous. It should have been incorporated years ago, for the benefit of people locked into these cities, that never get to see a field of wheat or a cow or anything. And reestablish the priorities with basic essentials in their proper place, you know."

The motivation for farm literacy programs was to increase awareness of the importance of the natural world in people's daily lives, including what they ate, wore, and other materials they used. A more implicit goal was to highlight the importance of family farmers in making these products available. AAM members were aware that their own futures as family farmers were increasingly bound to that of urban dwellers. They perceived their own numbers to be dwindling and their voice in issues of policy to be correspondingly weaker. The growing urbanization of Congress served as clear evidence of this reality.

We talked briefly about the Willie Nelson concert to be held at Texas Stadium the 14th of March (?). Willie Nelson has consistently brought attention to the problems farmers

are having and has raised money for them through his Farm Aid concerts. The Okla-
homa AAM has lots of ties to Willie Nelson. His daughter lives in Oklahoma and she is
involved. Also the hotline has received money from the Farm-Aid concerts. Mike Cops
(sp?) of CNN wanted to document what happens to the money—the OK people I think
were going to attest that money does indeed get to the farmer.

A kind of heated discussion then started when someone raised the issue of whether we de-
feat ourselves when we accept disaster relief payments. From what I understand: a
farmer chooses to enter this government program and abides by the government's limit
on how much land he can put into production—so that the government controls the sur-
plus. In turn, the government guarantees the farmer a certain price for that wheat per
bushel per acre—regardless if the land doesn't produce that much or if the price of wheat
falls below that amount.

On the one hand, those programs guarantee a price and some argued that it also pro-
duces stability within agriculture. Some people, on the other hand, argued that those
programs keep us from developing sound programs—which presumes that our present
course is not sound. That these programs are a type of welfare—"we are locked into a
modern day form of servitude," someone said. The main problem, this group argued, is
underconsumption.

They gave the example of foreign plants which can produce cheap stuff—because of rel-
atively low labor costs—thus increasing the profit margins of the companies. "Our peo-
ple are left out of work—so companies can make bigger profits."

Another heated discussion arose concerning state question #640 which would require a
vote of the people on any tax measure. Again, there seemed to be two sides, but it didn't
seem like a whole lot of people were committed either way. First of all, people were insis-
tent that we should take advantage of our right to vote. Some people thought the mea-
sure was good in that the government seemed to be "pretty free with our money." On the
other hand there was real awareness that people in the cities (Oklahoma City and Tulsa)
seemed to be for it. "When cities are for something, I get a little suspicious." A real sense
of powerlessness in some ways: that people in the cities outnumber us—that rural coun-
ties are drying up. Part of the people that were against the measure were saying that it
didn't include property taxes—an increase in property taxes (which people were imagin-
ing is what the state would have to do to make up for lost revenue) would devastate
farmers since they own so much land.

Fieldnotes, 7 March 1992

The AAM membership often discussed policy issues; the benefits and lia-
bilities of government programs were topics that engendered strong feel-

ings. This was true because government policy cut to the core of the beliefs of Oklahoma AAM members. During the meeting I describe above, one member made the point that government programs may help to promote a degree of stability within the agricultural sector, but at a price. Some felt that current agricultural programs diverted decision makers and farmers from attempts to develop policies that really helped farm families.

AAM members did not want subsidies. In fact, they felt that, given the low prices that they have historically received, it was the farmer that was subsidizing the American public's access to cheap food. Rather, AAM members were intent on getting a fair price for their product. Just like industrial workers who are able to organize and to obtain a living wage, members of the AAM felt that they too were entitled to commodity prices that not only covered their production costs, but also enabled them have a decent standard of living (Dudley 2000:151).

As Ritchie's (1979) interpretation of the CED reports indicates, strong anticorporate sentiments were held by AAM members. Many believed that government received its mandate from corporations, and, as a result, that public policy was aimed at curbing the progress of family farmers and, indeed, forcing their removal, while serving to further enrich food-related corporations. The increasing concentration of wealth into fewer and fewer hands was found to be contemptible. In addition, the corporations' constant global search for the lowest production costs to increase their profit margins caused members to rally against any global or regional free trade agreements. The abandonment of the American worker by corporations in their search for reduced low labor costs elsewhere was also deemed detestable by AAM members.

Mr. Gains commented, "And again, you get back to the government, we've exported all of our industry. You know we've got thirty some plants in Mexico building General Motors products, and that put a lot of people out of work. They move down there and hire those people cheap. And you know, you feel for those people . . . But here we put all of these people out of work. Then they bring their product back and sell it to you and I just like they'd given eight dollars an hour for our labor. A real interesting example the other day, and I saw something on TV today. These NIKE, is that how you pronounce it?"

"Ni-kee, yeah," I responded.

Mr. Gains looked at my feet. "You don't have any, thank goodness . . . fifty to a hundred fifty dollar a pair . . . they're built in Indonesia and this guy's contention was that they had about five dollars and sixty cents a pair in those shoes. Some guy working for fifty cents or a dollar an hour. I believe it was twenty-one cents labor in each shoe, forty-two cents in both shoes. They're shipped over here and they sell for that humongous price. And we got people out of work. . . . I had the opportunity to hear the discussion in our Congress in about '86 over that very thing. They were strongly trying to lift the rest of the sanctions against imported shoes at that time. And at that time, the number of people involved in the shoe industry in this country . . . for 70 million dollars they could pay them a living wage, you know, employment. [On the other hand, they could] lift all the restrictions and let the shoes come in here and they could save the consumer 250 million dollars you know. A real lucrative deal, but our poor people out of work and living on a meager income, so naturally [they] don't buy as much bread, as much rice, as much anything else. So the pile gets bigger and bigger."

Perhaps now, Mr. Gains's assertions in this chapter's opening conversation are more apparent. There he explains that farmers were not "in the boat"—this crisis—alone and that it was not overproduction that was the cause of depressed agricultural prices, but underconsumption. "It's those times when we've got a tilted situation and some major segment of the economy wasn't getting their fair share that the production started piling up." In the above example, the part of the economy that was not getting a fair share was workers whose industries had been exported. These unemployed workers, he reasons, are on meager incomes and thus cannot buy as many agricultural products, which, in turn, increases surpluses and drives down agricultural prices. According to the AAM theory, there is a triangulation of cause and effect that includes corporations, the government, and agriculture. They denounce corporations for their cupidity, their abandonment of American workers, and the exportation of industries. They blame government for failing to restrain corporations and protect American industries and workers. As a result, they argue, family farmers suffer.

There is a precedent for this anticorporate feeling in Oklahoma; it is a sentiment that emerges periodically throughout the state's history. Its origins can be traced to the experience of the pioneers. After the government arranged the non-native settlement of northwestern Oklahoma but before statehood in 1907, strong strains of radical political ideologies emerged among the settlers. Thompson (1986:221–222) states:

> Socialism, as well as populism and belligerent reformism, prospered in a frontier environment of which merchants, landlords and frontiersmen had not yet gained control. On the frontier, social, economic and political institutions were malleable, and pioneers were confident in their ability to create a new world.

5.1 Family with cow at their frontier home, northwestern Oklahoma.

Socialism was especially attractive in the most primitive frontier areas, where economic and political behavior had not been institutionalized. Moreover, the personalized nature of the system encouraged radical thought. The economic order in much of Oklahoma was so primitive that there often seemed to be no difference between a small businessman and a common criminal. Consequently, pioneer farmers and laborers often rejected reformism and considered revolutionary proposals.

These political beliefs were especially prevalent among people who had settled the new territory away from the cities and towns, that is, for the most part, farmers. At the core of such political sentiments was the recognition that the new territory had been settled and developed too rapidly and too ruthlessly. The swift development, Thompson (1986) argues, produced graft, overpopulation, overproduction, panics, and exploitation, and began to lay the foundation for one of the greatest ecological disasters in the history of the world—the Dust Bowl.

Fueling much of this haste and, consequently, the generation of radical ideologies, was the perception that corporations were largely responsible for this phenomenon. For instance, the *People's Voice* (a newpaper from Norman, Oklahoma), on 16 February 1895, commented that:

> If this republic ever falls it will fall through having too "strong" a government. The calling out of troops to put down labor trouble breeds contempt in the hearts of those who earn their bread in the sweat of their face. The past year has shown the people that the strong arm of government is all on the side of the moneycrats and grabocrats . . . the constitution and laws of our land are trampled under foot every day by corporations of this country; yet none of our big daily papers denounce them as "traitors" or "anarchists."
>
> (Thompson 1986:72)

Such beliefs were fairly widespread among the settlers. For example, the Populists in Oklahoma Territory were able to muster enough support to have a considerable voice in the territorial political structure. In the elections that were held during the early 1890s, Populist candidates were able to garner 18 percent of the vote in 1890, 21 percent in 1892, and 33 percent in 1894. This last election enabled them to win seven seats in the territorial house and five in the senate. By comparison, Democrats controlled only three in the house and one in the senate (Thompson 1986:72).

But as statehood approached, the influence of the populist People's Party began to wane and was eventually replaced by the Farmers' Union, the labor movement, and the Socialist party. What united all of these political constituencies was their belief in the need to control corporations and corporate monopolies, faith in the power of the common person, and a vision of a humane and cooperative society.

Many began to feel that significant reform within the territorial system would not be accomplished and they looked forward to statehood with the hope of having a hand in forging a more just society. To that end, many Populist, Socialist, and Democratic groups of farmers and laborers began to draft a constitution for the new state that manifested their political philosophies. They were inadvertently aided in their mission to have the resulting document accepted by the delegates to the Oklahoma Constitutional Convention by a series of corporate mishaps preceding the gathering. Nearly one hundred people were killed when a railroad bridge over the Cimarron River collapsed due to the negligence of the railroad company. Additionally, stories were circulating about businesses that were selling contaminated food products. Both incidents added to the growing public sentiment that "big business was an evil that required tight regulation in order to protect the interests of the general public" (Morgan, England and Humphreys 1991:46). The proposed constitution the alliance developed was ultimately adopted with only one significant revision.[5]

The constitution that emerged from the convention was a model of progressiveness for its time. The document called for strict corporate regulation, forbade trusts and monopolies, outlawed price discrimination, and mandated tax equity. In addition, it "eloquently affirmed humanitarian and pro-labor principles" (Thompson 1986:78).

From the perspective of the alliance that created it, the greatest achievement of the Oklahoma constitution was the restraints placed upon corporations. Article 9, the longest (longer than the entire U.S. Constitution), was reserved for corporate concerns. According to Thompson, the article required:

> [C]orporations to be chartered by Oklahoma, prohibited political contributions by corporations and required railroads to divest themselves of mining operations, reduce rates to two cents a mile and pay taxes on rolling stock and moveable property at the same rate as personal property. It then stipulated that these principles were to be enforced by a corporation commission whose members were to be elected by the people.
>
> (1986:79)

After the constitution was adopted and after Oklahoma became a state, the Democrats, who had portrayed themselves as reformists to capitalize on the growing progressive feeling in the state, held a very tenuous hold on political power. During this period, socialism continued to increase in popularity. The stronghold of socialism was, interestingly, in northwestern Oklahoma. Thompson documents the importance of settlers' experience with land in the development and maintenance of socialist beliefs. For instance, he demonstrates that the counties that had the lowest annual rainfalls, and thus the greatest challenges in production, had the greatest proportion of socialists (1986:128–129).

He also demonstrates that the traditional view of settlers doing battle with and ultimately subduing nature did not seem to hold true for socialists. Rather, they recognized that human labor had to work in concert with the cycles of nature to achieve the maximum benefit for both people and the environment. Again, part of this awareness arose from the experience of too rapid settlement and development. Two socialist farmers discuss these concerns in letters to the editor of a local newspaper (Thompson 1986:160):

> While arguing that the state should sell its school lands to the tenants, one Socialist farmer revealed his belief that "land is a tool of production" but that it is more: "It is mother of us all and while we may dispense with other tools we cannot with land." Another letter writer described land as a human "heritage" which human beings had a responsibility to protect for future generations.

The impact of the Socialist party in Oklahoma was of considerable consequence as Morgan and his coauthors indicate:

> The economically distressed citizens of Oklahoma were attracted to the Socialist party. In 1914 the party received nearly 20 percent of the vote in the gubernatorial election; a half-dozen state legislators and scores of local officials who ran on the party's ticket were elected. There were more registered Socialists in Oklahoma than in the state of New York. Thus a state that generally is viewed today as conservative registered the largest vote for a Socialist presidential candidate [Eugene V. Debs] of any state in American history. Socialist gains came at the expense of the Democrats, weakening the reformist wing of the Democratic party and providing the state with a third party nearly equal to the Republican party in voter strength. Clearly the Socialist party had inherited the Progressive agenda as it attempted to appeal to the most disadvantaged groups and their hatred of the "parasites of the electric-light towns."
>
> (1991:41)

Prior to World War I, however, persecution of Socialists in the state began as Democrats and Republicans wrangled for political power. Socialists were painted as unpatriotic and their ideology alien. Local defense councils, formed to bolster the war effort, often became the focal point of anti-Socialist sentiment. Their activities sometimes resembled that of angry mobs. These councils, unfortunately, also served as the training ground for the Klu Klux Klan, whose activities in the state increased during the 1920s (Morgan, England, and Humphreys 1991:49). By then, Thompson asserts, socialism in Oklahoma was all but extinct:

> It could thus be argued that Oklahoma's Socialists, who were veterans of the United States' last great frontier, recognized the nature of the Great Frontier. They saw that Oklahoma's sad history illustrated the crucial dynamics of capitalism's conquest of nature, and they rejected that process as barbaric. Instead,

they formulated an ideology which was the antithesis of the Great Frontier. Rather than an individualistic, materialistic, frantic and often brutal competition for wealth, Oklahoma Socialists advocated a patient, evolutionary political program designed to create a cooperative commonwealth consistent with the principles of Christian charity.

(1986:164)

Conscious of this history of agrarian radicalism in the state's past, I wondered whether the current members of the AAM were descendants of these socialists, populists, and labor activists. Why were they here? Was there something in these individuals' past that predisposed them to involvement? If they did not have personal and familial connections, I wondered whether they simply grew up hearing about these philosophies and ideas in the communities of their youth.

I investigated these connections with all of the farm couples of the study, but I simply could not establish any firm association. Though they told me that their parents were, by and large, Democrats, none of the men and women with whom I worked professed to have any personal or familial tie to ancestors who held radical agrarian ideas. Most had only a dim awareness of this revolutionist period in the region's past.

One potential explanation for this disconnection is the historical persecution of those who held alternative political philosophies to those tendered by Democrats and Republicans in the state. As the persecution of these individuals mounted, no doubt the stigma associated with these ideologies also increased. To keep themselves and their families safe, many men and women very likely kept quiet, or limited political and social commentary to small circles of trusted family and friends.

But to get back to the central question that motivated this line of inquiry: Why are AAM members present and active in northwestern Oklahoma? The cultural and historical information presented in this book presents one answer and, simply put, it is this: the structural, economic, and political forces that have periodically given rise to strong agrarian beliefs throughout the state's history are still present and vital.

We broke for lunch. After lunch we had our guest speakers who were two of the original founders of the national AAM.

These guys were a trip! They both looked like regular sort of farmer guys except they kind of had that wild-eyed look and I would swear, based on my professional experience in mental health, that one was on some sort of heavy duty psychotropic medication. Re-

member: there are two branches of the AAM; one is the group that believes in working through the political system—within the system. This is the branch that is active in Oklahoma. The other branch is what my informants call the "grassroots" wing and they do more demonstrations and I get the sense that they are not beyond violence. These two guys from [another state] were from this branch of the AAM.

They made this big presentation about how the constitution is the supreme law of the land (they handed out copies of the constitution). One of the guys took us through it and showed us where our rights were clearly spelled out. He then went on to show how the actions of government, especially in regards to the crisis, were compromising all of these basic rights.

They felt that what has given the government the right to do this is that they declared Emergency Rule with the banking crisis of 1933 and no one has ever repealed it. He says that because of this, people (farmers) have been placed in the role of servitude. For example he said, "Your land isn't really yours." Farmers, he claims, have a warranty deed. If they really owned it outright, they wouldn't have to keep paying taxes on it every year. Points also to the fact that government controls production and that grain is tied to the Federal Reserve in order to give it value.

I really thought people were buying it and they made it seem like one of these guys was some sort of hero. To be perfectly honest, I was pretty shocked by what they were saying and shocked that people seemed to be buying it. I really felt like I was at some sort of Klan meeting, especially when he started in heavy with the rhetoric: (1) "raised to believe in the Constitution, God, and flag," (2) "I fought for my country" (a lot of my informants have too by the way), and (3) "they're telling us what color people have to be to get jobs," "they're telling us when and where we can pray," and "they're letting homos teach in our public schools."

Fieldnotes, 7 March 1992

My strong reaction to the meeting was due to the realization I had afterwards of how truly rare it was for me to be frightened in my daily life—not anxious, not worried, not nervous, but truly frightened. It is how I felt as those men spoke. Being a minority myself on several counts, I remember trying to stay perfectly still and focused on my notes, hoping that I wouldn't call attention to myself in any way.

I did not have a language to desribe these men then, partly because the bombing of the Alfred P. Murrah Federal Building in Oklahoma City had not yet occurred. But I do now: these men were most likely militiamen. And Andrew—responsible for the disturbance in Washington—was one of the two individuals who spoke to us that day.

I was to have an interview session with Mr. and Mrs. Gains a few days after our AAM meeting in Woodward. I was not sure how I was going to broach the topic of the presentation by these two men, but they did it for me.

"Those two people you met there the other day . . . they're American Agriculture but that was, there's a split there. They're called the 'grass-roots.' They didn't want no structure, no leaders, no membership fees, nothing. And we splintered off from them in about '84," Mr. Gains said. "They don't help us fund anything and we work ourselves half to death just paying the salary of the three or four people that worked on the Washington [undecipherable]."

Mrs. Gains said, "And they don't . . . a lot of them don't pay taxes . . . and they believe in violence . . . fighting for what, I mean physically, for what . . ."

Mr. Gains jumped in, "But were dedicated people, and I don't know. What did you think about what he said [the other day]?"

I answered, "That's funny because I was interested in what you thought about it, because I was really surprised."

Mr. Gains replied, "With me, the magnitude of the way he expressed it, just blows my brain. I can't see, you know, that it could possibly be. I understand and have understood that they [the government] have that power and authority. It comes under emergency circumstances. But I've got to see whether the Emergency Powers of 1933 was officially released. Franklin Delano Roosevelt, when he implemented them, we had a definite emergency and he shut all the banks and stopped everything and reorganized and, when they opened up, the game went on. But I'm inclined to believe that [with] the new philosophy that appeared later on after World War II, they realized that they could control people more readily under these circumstances. The farm program is a typical example. They could control all economic factors more readily so they just left it in place. Whether they just deliberately left the emergency order in place and that's the reason for it, you know, I can't believe that."

Generally Mr. and Mrs. Gains acknowledged that they could not sympathize with the strategies of the grassroots AAM, but clearly felt respect for their depth of feeling and commitment to the cause.

Nationally, far-right and extremist organizations, some racist, some anti-Semitic, have waged a campaign for converts. These organizations have capitalized on the misfortunes of farm families—the displacement of people from their land, unemployment, financial distress and hopelessness—and recruited them for membership. They have had some limited

success. Some of these organizations believe in a "constitutional funda-
mentalism," which essentially means that the United States is not right-
fully a democracy but a Christian republic, that state and federal laws developed by legislatures are not lawful, and that Christian common law
should reign supreme (Levitas and Zeskind 1987:24). Others have focused
their efforts on the Federal Reserve System, whose work they believe has
been responsible for the displacement of farm families.[6] They blame a
shadowy "international Jewish conspiracy" for this.

The Posse Comitatus (meaning, "the power of the county") is one such
extremist organization. Its members acknowledge no authority higher
than the county sheriff and believe that Americans are legally bound to ob-
serve only the first twelve amendments of the U.S. Constitution. They
further insist that the federal government is largely in the hands of Zionist
conspirators. As a result, they view the observance of any national laws or
regulations on the part of citizens, such as paying taxes, as complicitous
behavior. To put a stop to this conspiracy, the leaders of the Posse Comitatus advocate vigilantism (Stock 1996:171).

Disenfranchisement and despair have led many to join these groups in
search of direction and a sense of hope. Stock argues that perhaps these organizations served another function. For many men—"where farming and
industry offered dead ends, incomes were stagnating at best, farm organizations had been either demasculinized or feminized, and the promise of a
good life was practically a joke"—these organizations may have offered "a
chance to prove their masculinity and honor" (Stock 1996:170).[7] Finally,
these organizations provided a framework to individuals to help them understand their situation, for how could farmers, who were doing so well in
the 1970s, all of sudden experience such hardship unless some government
conspiracy was involved?

Given that these extremist right-wing organizations aimed to meet important needs for potential members, what is most compelling about the
farm politics of the 1980s to Summers (2001) is that they were not *more*
successful. She attributes their limited achievements to the broader vision
for social change that family farm activists promoted. Their vision often
went beyond agriculture, as I believe Mr. and Mrs. Gains's statements indicate, and included a concern for the environment, just trading policies,
and the welfare of workers. The ability of farm organizers to link with
other social activists, increased their political strength. The result was,
Summers notes, that left-leaning interests came to dominate policy reform
efforts. Provocatively, she goes on to suggest that the "countryside was in
fact one of the few places in the 1980s where progressive activists seriously

engaged and combated right-wing ideologies" (Summers 2001:317) of the Reagan era.

In an effort to do just that, many mainstream religious organizations banded together to provide education and training to family farm organizations in an effort to stall the efforts of extremist hate groups.

Mr. Gains said, "And I know I'm on everybody's list and they watch us. We've had so many fears along the way that we was going to fall prey to some racist bunch, and Nazis and what have you . . ."

I responded, "You mentioned that. Why is that? Why are they afraid that American Ag would become full of racists?"

"Well, psychologists must feel like that the predicament that farmers are in [makes them] more susceptible to follow somebody with these racist tendencies or something. You know . . . the Jewish people's fault, that they're the ones that caused all this . . . and there's people that believe it!"

Mrs. Gains said, "We went to this meeting for two days in Guthrie, wasn't it?"

Mr. Gains responded, "Yeah . . . They had a representative of the Jewish people and the Negro and the American Indians. . ."

"And Prairiefire [an advocacy organization based in Iowa that works on behalf of farm families in the Midwest]," Mrs. Gains added.

I asked them who organized the event.

Mr. Gains responded, "Well, it was through the Conference of Churches and the Black [undecipherable]."

Referring to extremist organizations, he added, "They have tried to infiltrate us."

Surprised, I asked, "They have tried? How have they tried?"

Mr. Gains responded, "Yeah . . . well, just on two occasions. One was this young man down here in the southern part of the state . . . didn't make no bones about it."

Mrs. Gains added, "He's a Heritage . . ."

"Yeah . . . racist of some kind . . . but he never made no headway at all . . ."

"No, in fact, he sent us a tape and we put that tape on and it never, it didn't have a thing on it. Just one little bit of him sitting under a tree or something," Mrs. Gains said.

Mr. Gains added, "There was something fouled up and he didn't know it or something, and so I always wondered what was on that. But yeah, we had to make a written statement and it's still on file down there, that we have no part of any racist organization of any kind. And we have had to expel one or two of our members because they have tendencies to be more militant . . . [Mrs. Perkins] had trouble with [G——], 'cause he was, he certainly had tendencies of being racist, and of course, his big beef was with the Federal Reserve System."

As I previously mentioned, Mr. and Mrs. Gains did not support some strategies, particularly violence, that some more radical individuals and groups utilized to advance the cause of family farming, but they certainly respected the depth of feeling and commitment that many of these activists exhibited. The respect the Gainses felt for anyone who held a strong commitment to agrarianism was partly due to their belief that their way of life was being threatened. They likened their situation to that of an endangered species. I often heard the phrase, "a vanishing breed."

Mr. Gains told me, "You know they make all kinds of commotion over the spotted owl that's disappearing, or the snail-darter fish, or white rhino or something, which is a shame. I don't want that to happen, but [will they] let a breed of people disappear from the face of this earth that have invested tremendously in this country?"

Implicit in this notion of a "vanishing breed," was the perception that they were one of a kind. In response to my inquiries about the nature of the farmer, AAM members often told me that what a farmer was or what made him "tick" was a mystery. I might ask, for instance, "What is a farmer?" or "Would you define 'farmer' for me?"

Mr. Gains, for example, responded, "I have been trying to figure that out myself and I just don't know."

Similarly, Mrs. Gains said, "I don't know what makes them tick. I have lived with him long enough to know. You cannot define a farmer. You cannot."

Underlying many of the activities and beliefs of the AAM membership is an explicit dichotomy that positions industrial understandings of agriculture and rural life on one end of a continuum and their own beliefs about the same on the other. The first is scientific, codified, structured, and defined. The AAM vision, alternatively, is enshrouded in what I came to call the "unknowability" of farming: farmers are unique and impenetrable and their knowledge or "know-how" mysterious and inscrutable. If profit margins, management science, production strategies, and efficiency compose the core beliefs of industrial agriculture, then words like family, kinship, community, tradition, vocation, and stewardship—notions that resist ready quantification—may capture much of the essence of family farming. The foundation of such a vision of farming and rural life, I would argue, can be found in the religious beliefs of my informants.

Of God, Men and Farming

Because of her connection with AG-LINK, Mrs. Perkins was one of my first informants. During the course of our initial conversation, I asked her about potential subjects for this research, individuals I might interview who had experienced, or who were now experiencing, severe financial distress. She telephoned a few farmers to obtain their permission to share their contact information with me. When she gave me the list of names, she made the point of telling me, that these were "good solid, Christian citizens." I was later to learn that the people she suggested I interview were all members of the AAM.

Similarly, Mrs. Ross offered to have a social gathering at her house, to introduce me to members of the AAM and to give me the opportunity to fix interview appointments with potential informants. The event was a great success. I record in my fieldnotes that Mrs. Ross told me that one of the things that AAM members share is that they are all religious.

Both conversations struck me as curious, and I realized then that Mrs. Perkins and Mrs. Ross were communicating to me in "Oklahoma code." This was their way of letting me know that AAM members are not extremists, but mainstream, or at least justified, in their thinking. But when I began to interview farm men and women and learn more about the AAM, I realized that these attributions of religion were more culturally significant than I originally thought.

Family farmers told me that, unlike corporations, they were concerned about land and took seriously God's command to till the soil and care about creation. Many members of the AAM spoke explicitly about the relationship between farming and religion; it was one of the discourses from which farmers drew meaning and it was appealed to by them to help me comprehend their position. Here, I explore this relationship, particularly as a justification for the agrarian vision of rural life, a life they felt was essentially moral and good.

I support Behar's (1995:6) assertion that "feminist revision is always about a new way of looking at all categories, not just at 'woman.'" Delaney's work (1991; 1998) exemplifies such an approach. She takes a broad and radical perspective by going straight to the source, examining origin myths, interrogating the values implicit in such stories, and elucidating their consequences for us today in terms of social organization, ideas about gender, and the more expansive cosmological and metaphysical views they support.

For example, in her ethnography of a Turkish village, Delaney (1991) examines residents' conceptions of procreation and demonstrates how their ideas are intimately and symbolically tied to origin myths. Through her analysis of village life, including the social meanings associated with land, marriage, and household and community structure, she demonstrates the social legacy of the particular human notion of procreation known as monogenesis and its relationship to monotheism and patriarchy. Her argument, which is continued in her provocative examination of the biblical account of Abraham in *Abraham on Trial* (1998), is important to us because it will help to further elucidate the nature of the relationship between men, land, and religious belief. But before we examine the foundation of this linkage, I believe it is important to briefly review the story of Abraham.

In this narrative, found in the Hebrew Bible (Gen. 22), God asks Abraham to take his son, Isaac, "whom thou lovest" (Gen. 22:2), to the land of Moriah and to present him there as a burnt offering. Carrying the wood, fire, and knife, they near the designated location after three days of travel. Unaware of Abraham's intentions, Isaac asks his father about the whereabouts of the lamb that is to be sacrificed. Abraham responds, "My son, God will provide himself a lamb for a burnt offering: so they went both of them together" (Gen. 22:8). When they arrive at the specific site, Abraham builds an altar, lays the wood upon it, binds his son, and places him on the altar. As Abraham prepares to slay his son with a knife, he hears the angel of the Lord call out of heaven, "Lay not thine hand upon the lad, neither do thou anything unto him: for now I know that thou fearest God, seeing thou hast not withheld thy own son, thine only son from me" (Gen. 22:12). For his unquestioned obedience, God later pronounces his reward upon Abraham: "That in blessing I will bless thee, and in multiplying I will multiply thy seed as the stars of heaven, and as the sand which is upon the seashore; and thy seed shall possess the gate of his enemies; and in thy seed shall all the nations of the earth be blessed because thou has obeyed my voice" (Gen. 22:17–18).

And indeed, because of his willingness to obey God even at the expense of his son's life, Abraham's actions have become the model of faith, and Abraham himself, the *father* of three major monotheistic religious traditions: Judaism, Christianity, and Islam. But it is this very notion of fatherhood that Delaney critically examines. Why, she asks, did Abraham assume the child was his to sacrifice? Would the omniscient God really only ask "one parent for the child, knowing that a child belongs to both mother and father or, perhaps, to neither?" (Delaney 1998:7) The story clearly conveys the impression that Isaac belonged to Abraham in a way he did not belong to Abraham's wife, Sarah. Pervading Genesis, and reflected by the story of Abraham is the notion of a powerful fatherhood—but is this notion natural or a consequence of cultural ideas related to procreation?

Delaney argues that monotheism is inexorably linked to the notion of monogenetic procreation. In both monotheism and monogenesis the generative and primary creative force is understood as coming from one source. In the divine sense, as discerned by the Abrahamic traditions (Judaism, Christianity, and Islam), the outcome of the creative process is creation itself and its one source is God. In the human sense, the result of procreation is children and, Delaney argues, that the Abraham story, other biblical literature as well as religious traditions have supported an understanding of the sole source of their genesis as being male.

Symbolically then, monogenesis is bound to monotheism, and the life-giving capabilities implied by each associate men with the divine, or God. In other words, men, through procreation, serve as a conduit of the divine, a notion that is an integral part of the definition of masculinity. Delaney (1998:34) states that, "procreation has been imagined as the vehicle for channeling divine creativity to earth. And ever since Abraham, it is men who embody the power to do so. That is the basis of patriarchy."

This book has discussed a number of cultural practices that undergird patriarchy within farming. Among these are inheritance patterns that favor men in the acquisition of land, the desire for land to remain in the patrilineal family name through the generations, and cultural beliefs and social institutions that support men as the primary agricultural decision makers—that is, as farmers. These and other practices work in tandem to "naturalize" men's identification with land.

You will recall from chapter 3 my strong desire to position land in the kinship charts of families—so powerful was the language they used to symbolically bind land to their families. One issue that emerged in the section on kinship and land was that farming was viewed as a generative and creative process by the men with whom I worked. In fact, when I explored kinship ties to land, farmers said that land most resembled a child. They said that farming land was a creative endeavor and that, like children, land reflected the amount of work one put into it. Also, they said that this relationship encompassed years-long passages of time.

Inspired by Delaney (1991), who found a symbolic association between women and land, since both serve as the nutritive material in which seeds are placed by men, I asked my own informants about their perceptions of land as gendered—as either male or female. The men and women who participated in this study, however, were perplexed by this question. None had ever thought of land as having a gender. When I pressed and asked about the gender of the child they said that land resembled, again, they were confounded.

My point is that men—not women—were seen as the singular creative force that imparted land its identity as a member of the family, or simply as a successful and fruitful piece of ground. Men, because of their definition as the farmers, as the critical decision makers, were thought to give land their essence. Whether it was a productive piece of land was considered men's doing—despite any amount of work contributed by women. Land was linked to men and related to them in a way in which it was not related to women. Being able to nurture land and successfully raise crops was similar—and in some ways was *equivalent*—to being able to successfully raise children, and like children, land was one of the criteria upon which men would be judged by the community.

The association between men as genitors and God as creator was made explicit to me by an Oklahoma rabbi. My informant said that God planted a garden in Eden. He created men in his image and his first command was to till the garden and to care for it (Gen. 2:15–16)—a point my informants often brought to my attention. Further, Rabbi Marc Fitzerman said that this commandment positioned man as a tool or instrument that God could use to implement his own plan of perfecting and caring for creation. Significantly, my informant added, that to fail as a farmer, was to fail utterly to do what God intended, since farming is the preeminently human—which is to say, male—task.

As already mentioned, symbolically, according to Delaney's argument (1991; 1998), women are viewed as the nutritive material in which the procreative force or seed is placed and that allows the seed to grow. Perhaps this is why women were seen as such an important part of the family-oriented farm operation, in which there was often a fusion of the public and private spheres. Women were considered vital to the fulfillment of goals: keeping land within the family and transmitting it to the next generation. Because of this patriarchal foundation, the devaluation of reproduction and domesticity was resisted.

Additionally, farm life was praised particularly for the values and lessons it enabled parents to impart to their children. Couples explained that children learned skills and were given chores beginning at an early age. The successful functioning of the farm enterprise depended on everyone doing their part. If a chore was not completed, other family memebers were affected. Children learned that they were an essential and valuable part of the family and farm. Farm life also presented a natural laboratory for children to learn about biology, sexuality, conservation, and the environment. It was clearly a way that agrarian values concerning the care of and responsibility for land were transmitted to the next generation.

The biblical passage that my rabbi informant referred to in our conversation included not only the command that men till the soil, but that they care for it as well. The notion of stewardship arose often in my interview sessions with members of the AAM as well as other family farmers, sometimes unexpectedly.

Mr. and Mrs. Cobb had been talking about banking practices when Mr. Cobb said, "Yeah, and it's funny when you go to the bank. I've had bankers tell me, 'Well, I'd take that guy as a customer in a minute because he's got lots of land.' I told the banker, 'Just because he's got lots of land don't make him a damn good farmer.' I said, 'He could have got that land by inheritance, or he could have got it by screwing somebody out of it. 'Cause he's got a lot of land don't mean he's a good manager and paid for all of it.' So, you know, land is pretty good collateral, but what's really funny is people think they own their own land. They don't own their land; they just have a piece of paper that says nobody else can take it away from them until they sell it."

Mrs. Cobb said, "It's just yours as long as you're alive. It belongs to God."

More predictably, when I spoke to Mr. and Mrs. Becker about erosion and soil conservation, Mr. Becker said, "I used to not [want to] get on a soapbox about caring for the land, but since I got involved with conservation programs, I've found that they [conservation programs] are necessary. There's something that we really need and should have."

I responded, "And I was surprised to find that I've heard that from several farmers. I guess I didn't expect to. And I have heard the phrase 'stewards of the land' more often. I mean, if I had a dollar for every time! I've heard that phrase a lot."

Mr. Becker said, "If you are a religious person, you have. And then [if] you're a farmer you learn this, if you grew up in a religious, biblical home. This is part of the Bible. This is something God commanded us to do. To take care of the land. To not waste it. Stewardship is important. Stewardship is important in whatever we do to be a good person, to be a good Christian, of whatever persuasion. There's no excuse for not being stewards for what has been provided for you."

Mrs. Becker added, "Yeah, it's a gift from God. The land and the world is a gift from God, so we should take care of it."

These comments linking farm couples' understanding of land to religious sources should not be a surprise.[8] Land is a central metaphor in both the Hebrew and Christian Bibles. The books of the Pentateuch, especially Leviticus and Deuteronomy, provide an understanding of land as a base of the covenantal relationship between God and his people (unknown author, n.d.). As I have discussed, according to Judeo-Christian tradition, land is held custodially; it is owned by God. Obedience to God determines whether one or one's group holds onto land or loses it. In the Hebrew Bible, for example, Jews came to inherit the land of Canaan because they were deemed worthy but eventually were removed by dint of their own sinfulness.

The idea of land as the basis of a covenant is one that carries into the Christian Bible. In the fourth and fifth chapter of the Acts of the Apostles, the story of Ananias and Sapphira sends a clear message that the relationship to the land is fundamental to the viability of the community of faith (unknown author, n.d.). In this particular story, we find a young Christian community struggling for survival despite foreign economic and political challenges. The members of the community have committed themselves to one another: they practice communal ownership and any money they receive for the sale of individual property or possessions is laid at the feet of the apostles. Each person receives resources according to his needs.

However, one couple opts for their own advancement. Ananias and his wife Sapphira sell a house, and turn over part of the proceeds to the apostles, but withhold a portion to further their own interests. Ananias is discovered and severely chided by the apostle Peter, who tells him:

> Ananias, how was it that Satan so possessed your mind that you lied to the Holy Spirit, and kept back part of the price of the land? While it remained, did it not remain yours? When it was turned into money, was it not still at your own disposal? What made you think of doing this thing? You have lied not to men but to God.
>
> (Acts 5:3–5)

When Ananias hears the scathing words of Peter, "he dropped dead, and all the others who heard were awestruck" (Acts 5:6). A few hours later, his wife tries the same trick and is again challenged by Peter:

> "Why did you conspire to put the Spirit of the Lord to the test? Hark! There at the door are the footsteps of those who buried your husband; and they will carry you away." And suddenly she dropped dead at his feet.
>
> (Acts 5:9–10)

These examples, from the Hebrew and Christian Bibles, are about the intervention of God when the values and beliefs of the community have been ignored to advance the interests of individuals (unknown author, n.d.).

What unites the various points raised in this section is that faith and indeed everyday life for many of my informants is influenced by religious discourses and texts. It became apparent to me that the Bible and many of its themes resonate in rural peoples' lives in a way that they may not for city dwellers. Like the agricultural societies in which the sacred texts for the Abrahamic traditions were written, in rural Oklahoma, people's lives are centered on seasons and their growth cycles. One informant, a displaced farmer now living in the city, told me that he is often called upon in his men's faith group to interpret biblical stories involving animals and agriculture.

I would argue that the symbolic connections I have outlined are the reasons I was often told by the subjects of this research that the family was the "natural" locus for the work of farming—not corporations—and why they believed that moral goodness inhered to the lifestyle. Considering this context of farming, another positioning of men for crisis may involve religion: to fail as a farmer is to have failed at their divinely ordained task. As a consequence, they are deprived of key expressive opportunities to demonstrate their faith and to live out central metaphors of their religious beliefs.

Creating Community

As I mentioned previously, I conducted research in "pockets" of communities. For instance, I might have two to four couples and other informants within a ten square mile area. My informants typically knew each other, but none knew with whom I was working. On more than one occasion, I was asked by one member of the couple if they were the only ones experiencing financial crisis. One farm woman, near tears, asked me this question, and though I had interviewed three of her four neighbors and had learned they were all also living under heavy financial burdens, she sincerely had no idea that anyone she knew was having problems. She clearly felt isolated. In response, I assured her that her family was not the only one and I encouraged her to reach out to others. This moving experience threw into stark relief the anomic structure of rural areas and the isolation of nuclear families.

A central concern of the AAM is that the "landed interests of individuals" not take precedence over the concerns of the community. In other words, this organization was focused on the conscious formation of community, forging meaningful ties in a culture whose social relations, because of the intensification of community capitalist practices and the endorsement of industrial values, have become much more diffuse (Dudley 2003).

When interviewing members of the AAM, I was struck by the language they used to describe their relationships to one another. It was the language of kinship. They often talked about one another as "family." These bonds seemed to be present among AAM members who felt they shared a commitment to family, kinship, community, and an overall agrarian perspective. And even though their bonds to one another were not based on blood (one of the bases of the American folk belief system for demarcating kinship relations; see Schneider 1968; Yanagisako and Collier 1987; Yanagisako and Delaney 1995; and Weston 1998), they shared a reverence for the continuity of the biological tie to land.

These family bonds were solidified by working together to achieve the organization's goals: attending meetings as well as regional and national conferences, working with the state agricultural hotline as first responders, and participating in political advocacy projects. These bonds were made manifest during periodic large social gatherings held on an individual family's farm. I attended several of these and they indeed had the feel of family reunions. The familial tie was not limited to social or professional contact, however.

I asked Mr. Gains, "Do y'all help each other out financially, like family members do?"

He responded, "In lots of cases, yes."

"Like in what way?" I asked.

Mr. Gains responded, "Well, I know of one individual that's older than us by fifteen years . . . His wealth came from assets that he accumulated all his life, antique stuff. And he sold one of those pieces of stuff for a large amount of money and I know of two neighbors that he helped out when they was going under . . . He sold this old asset, an old car . . . why evidently it must have been sixty, seventy thousand dollars worth, that he didn't think they deserved to go that route and he paid them out enough to where they could get along. And there have been other examples."

The tie of community, if not kinship, was also not limited to people in close geographic proximity. Mr. and Mrs. Gains talked about AAM members from across the country—some they had never met—who, when traveling through the state, would stop in and stay with them. The presumption being that membership in the AAM was enough to create a solid bond of mutual understanding and trust. Speaking of his relationships to people within the AAM who have been long-term participants, Mr. Gains said,

"For those few I, gosh, we've just all become family. That don't stop within the state. There's people we've met all over this country. It's a real good feeling."

Becoming a member of the AAM provides an opportunity for individuals to informally share their personal stories of crisis, an experience that to some degree legitimizes membership and helps to create and strengthen the bonds of community. These stories are sometimes tragic and are filled with emotional content. Many members have dealt with their own financial crises and, through their own experiences, have learned that sharing their feelings is a good way to cope and heal. They encourage others to do just that. In the process of forging community, they also challenge the proscription against the expression of emotions by men.

I asked Mr. Gains what he and other American Ag members talked about when they got together. He gave me a list of topics, many of which I have discussed in this chapter. He ended his statement, however, by saying that there is "some degree of visiting or discussing some particular person's problems or how to help them, try to help them."

I asked him, "Is it okay to discuss problems?"

His response betrays a concern with the isolation of rural areas. He said, "Yeah, when anybody finds them[selves] with difficulties, they feel free to talk about it. This is amazing because, again, in the community like this to go up to the little town in the morning to the Quick Stop and to say, 'Hey, I've been sued' or 'I'm about to go broke' or 'I'm going to have to file bankruptcy,' you never know about it until after the fact—you never know nothing about it—and that's pride I guess and . . . but that's something, even in church it don't happen. Have had several people that got to go completely broke and be sold out that belong to the [local] church up here that has a pretty good congregation and nobody knew about it until it came down to the wire."

"Why not in church you think?" I asked.

"I've tried and everybody else has tried to get them involved . . . but nobody shows up from out there. Sometimes you get the feeling like you got some mental thing wrong with you or some dreaded disease or something or other like the plague."

"If you have financial trouble?"

Mr. Gains responded, "Well if you talk about this stuff—even talk about it. Yet a lot of these people as individuals know the possibilities, knows it's there, but they don't . . . well a lot of them will make a remark, 'Well it's going to happen but you ain't going to stop it.'"

I asked, "So you feel like maybe you have a sense of community with the American Ag Movement that you don't have with your neighbors?"

Mr. Gains responded, "Right, oh absolutely."

Mrs. Gains joined in, "That's right. Absolutely, *absolutely!*"

Mr. Gains continued, "And it's that way in any little given area. Those that belong and really stayed tough and were there, showed up all the time, helped all they could, physically and financially—there's a closeness that you can't describe. You know we've spilled our guts to each other with our personal business, our spiritual feelings, cried on each other's shoulders. Just close, like [Mrs. Perkins], you know, she's just one of the family that's all."

When I asked him what he felt united them as a family, Mr. Gains responded, "Well, it's those common beliefs, I think mainly is the most binding thing that, like I said, they have so much respect and consideration for the blessing of the soil and the earth, and they're rooted so deeply into it and they feel it's so basic, so fundamental, so vital, and so blessed. And I think more so today, there are so few of us, they feel fortunate to just be out here. And they feel an obligation to leave it as good or better than they found it, and this strong desire to perpetuate something that has contributed so much to this country and the whole world really . . ."

Mr. Gains told me, "But that was the goal, that we had to succeed and then hand our operation on to one or more of those boys, and like I said, the American Agriculture Movement, that's basically what all those people are all about. It means a great deal to us to have to terminate something that went through three and the start of four generations.

"I told you the other day, I feel like if there's a heaven, my dad's there. And if he had any preference, he's right there where he was born and raised. He loved it and I think that goes for any old person his age or older that I ever knew. I think their spirit draws right there, to the places they loved, that was good to them. And they tried to establish it so it would stay and last. If you don't do your part to preserve it and perpetuate it, then you're not upholding that tradition.

"And anybody, most any of the older ones that you talk to, are aware of the investments granddads and dads, parents and grandparents made, but yet they don't show me that they're willing to defend that. We've got two little cemeteries here that's full of old people that settled this country. To me, out of respect to them and the love they had for this, I'll defend it till I'm dead. But we ain't got very many that will. And why is that? Something that was established and built on, and built on, and built on, and then just let it die?"

Mrs. Gains added, "They talk about pride and people having too much pride, but I don't know sometimes if you can have too much pride in things like that."

Tom laughed uneasily, "Well, maybe like Casey says, a fella ain't got a soul of his own, but on'y a piece of a big one—an' then—"

"Then what, Tom?"

"Then it don' matter. Then I'll be all aroun' in the dark. I'll be ever'where—wherever you look. Wherever they's a fight so hungry people can eat, I'll be there. Wherever they's a cop beatin' up a guy, I'll be there. If Casey knowed, why, I'll be in the way guys yell when they're mad an'—I'll be in the way kids laugh when they're hungry an' they know supper's ready. An' when our folks eat the stuff they raise an' live in the houses they build— why, I'll be there."

—John Steinbeck, *The Grapes of Wrath*

Speed now the day when the plains, the hills and all the wealth thereof shall be the people's own, and free men shall not live as tenants of men on the earth which Thou has given to all. Enable us humbly and reverently, with clean hands and hearts to prepare ourselves for the day when we shall be Thy tenants alone and help us become faithful keepers of one another and of Thy good earth—our home. Amen.

—*Ceremony of the Land*, Reverend Howard Kester and Evelyn Smith (Munro)

What the social world has made, the social world, armed with knowledge, can undo.

—Pierre Bourdieu

CONCLUSION

Modernity, Emotions, and Social Change

The Broken System

Sloan (1996) addresses the impact that modernity and the capitalist industrialization and bureaucratization that has accompanied it has had on the psyche. Employing Habermas's notions of the "lifeworld" (the collective cultural knowledge that serves symbolic functions) and "system" (knowledge and activities that relate to the material reproduction of a society), Sloan explains that with capitalist expansion, cognitive-instrumental rationalities (the system) increasingly extend themselves into the spheres of the lifeworld so as to ensure efficient social administration and successful market operations. These instrumental technologies disrupt traditional and symbolic communicative activities. As a result, there is a growing chasm be-

tween individuals and the communicative and community-building activities that have customarily provided the basis for the construction of social
meaning. In other words, individuals are left to their own devices or with
only the logic and values of the system to find personal meaning and create
identity. Since the private realm "has been stripped of the contexts of cultural and material reproduction," Sloan argues, "the individual comes up
empty-handed—or with a handful of prefabricated dreams." Habermas, it
is interesting to note, refers to this process as "colonization" (see Sloan
1996:47–66).

The notion of colonization was on my mind during fieldwork for several reasons, though it is a concept that I use with reservations. Despite the
diversity of agricultural goals and strategies of farm families discussed in
this book—whether "family-centered," "industrial," "yeoman," or "entrepreneur"—it is clear that all farmers are aware that they are producing for
capitalist markets. Thus, I am not suggesting that my informants, particularly the members of the AAM, have some claim to a putatively more satis-
fying moral order *before* capitalism that was then "colonized" (Jane Adams,
personal communication, 2 July 2002). Sometimes reluctantly and sometimes enthusiastically, the farm men and women of this study participated
in the increasing modernization of the agricultural sector and rural life in
northwestern Oklahoma. But it is also important to remember that there
are many "capitalist logics" and many capitalisms (Yanagisako 2002).

The men and women with whom I worked, were not opposed to capitalism, but to its contemporary practice in the agricultural sector. Indeed
my informants experienced an intensification of capitalism in the 1970s and
1980s, which increasingly stripped away any social or cultural factor that
could not be represented on a balance sheet. It was a type of capitalist, agricultural practice that was more quantified, codified, and mechanized and
in whose discourses could be increasingly discerned the "universal" language of science: essential truths that can be measured and geographically
transplanted with little concern for local cultural contexts. In this sense it is
appropriate to use Habermas's term colonization to define the region's increasing involvement with broader national and international processes
that enabled and directed local cultural change: the transformation of rural
social relationships and institutions, the devaluation of certain practices and
subjectivities, and the search for individual and social meaning.

For example, the concept of "farmer" was itself being colonized or
transformed, linked now much more to management science and rationality and less to family, community, and the notion of stewardship. Reflecting this change, it seemed to me that farm magazine articles focused on ef-

ficiency and strategic thinking to ensure profitable decision making. In addition, while they have always concentrated their efforts on the augmentation of productivity, yields, and profits, the goals of research at land-grant universities are increasingly being defined by transnational corporations, which diverts scholars from alternative topics of investigation and the promotion of diverse forms of production and profitability such as "localization"—the production of food and fiber products aimed at local or regional seasonal markets.

The notion of "community" was also being altered through technology. Technological innovation has almost eliminated the need for the exchange of agricultural labor. Television has undermined the motivation for the organization of social gatherings. Competition, an integral component of the economic logic of expansion, has fueled an already strong sense of individualism—further isolating nuclear families. This competition also promotes a change in cultural values associated with land: it is being stripped of its symbolic power and its bond to kinship is loosening. Land is becoming just another commodity that can be bought and sold.

Also the notion of the colonization of northwestern Oklahoma resonated with reading I have done in the field of development. It was apparent that rural Oklahoma was finding itself more and more in a relationship of dependency to the metropoles. For example, agricultural policies are being developed in Washington and other urban centers that are geographically and morally distant from the world that most of my informants inhabit. The voices of family farmers, willing and eager to contribute to policy debates, are being drowned out by the moneyed chorus of stakeholders in the complex process that is agricultural policy development. As a result, the men and women with whom I worked have little say, in practical terms, about their own futures.

Finally, I was witnessing colonization in its most classic sense: the relationship of rural Oklahoma to the metropoles/core is essentially an extractive one. The area exports its commodities, natural resources, profits, services and, most importantly, its educated youth. Too far from urban centers to attract commuters and too barren, some might feel, to attract retirees, northwestern Oklahoma has not benefited from the fairly recent revitalization that has taken hold in many rural areas (Salamon 2003). Instead, according to Fitchen (1991), many rural areas are witnessing the introduction of drug trafficking and abuse, prisons, land fills for urban waste, toxic waste dumps, large franchises, and poverty, among other phenomena. Some of these already have a significant presence in rural Oklahoma.

Many of my informants believed that this process of transformation of rural communities they were experiencing were being driven by a handful of transnational corporations—corporations that control commodities and the entire production and distribution process, from seed to shelf. They believed their way of life was being sacrificed for profit maximization. And indeed agricultural corporations make huge profits: while 600,000 families went off the farm in the 1980s, food companies were receiving a 20 percent to 25 percent return on their investments—second only to pharmaceutical firms (Heffernan 1991).

My informants recognized that they were not alone and that the economic transformations they were experiencing were global in nature. Many were knowledgeable about and sensitive to demonstrations by farmers and other workers around the world. Farmers here and abroad, they told me, have been especially hard hit as governments wrestle with globalization and make difficult choices about subsidy programs that have supported their agricultural industries and workers.

Globalization—this process of capitalist expansion across national boundaries—is driven both by the imperatives of the market and by the actions of policy makers. For the last two decades the policies of the G8, whose members include the leaders of the world's largest economies, have been directed toward the creation of a unified global economy. Trade accords, incentives and disincentives of public institutions, such as the International Monetary Fund (IMF), the World Bank, and the World Trade Organization (WTO), have promoted the removal of barriers between national economies in an effort to make the arena of economic competition a level playing field (Sweeny 1999:1).

But the process, obviously, is not complete and it isn't a level playing field yet. Globalization is promoting what AFL-CIO president John Sweeny (1999) calls a "race to the bottom," in which countries compete with one another for transnational corporations interested in the cheapest production costs possible. In the name of such competition, governments (including ours) seem all too willing to ignore or dismantle policies that would protect their own people. We are periodically reminded of this reality via media reports regarding the injury of workers, the exploitation of women and children, and human trafficking.

But is this race to the bottom inevitable? I am continually surprised to the degree that many of my informants, colleagues, and friends alike view capitalist expansion as a "natural" process. Many farmers felt that not much could be done to turn the tide of corporations' involvement in agriculture, given the government's seeming refusal to put reasonable limits on their activities. Also, they were convinced that they would be the last generation to reap the benefits of rural living and agricultural work. It seems to me that corporations and our government have done an excellent job of portraying capitalist expansion as an expression of human nature, one that fits especially well with the individualism of American culture. Particularly during the Cold War, U.S. foreign policy and government information agencies were focused on the message of communism's ineluctable failure (Belmonte 2003): human nature—the will to compete, to be better than, to have more than—will win out in the end. It is a powerful message that many internalized.

But to many of my informants this unrestrained globalization did not seem natural nor did it seem right. During the course of my research, a few men and women affiliated with the American Agriculture Movement raised the religious notion of "jubilee" as a biblically proclaimed ideal of a just society—a notion with which I was not familiar. The idea of jubilee is contained in Leviticus 25:1–55. Jubilee follows seven "weeks of years" and thus occurs every fiftieth year. The jubilee year has three key features: (1) the land is to lie fallow, (2) all landed property which has been sold is to revert to its original owners, the families that received it when the land was apportioned by lot after the conquest of the Canaanites by the Israelites (Josh. 13–21), and (3) Hebrew slaves are to receive their liberty (Rabbi Marc Boone Fitzerman, personal communication, 7 February 1992).

The intent of the jubilee year is explicit. The land is God's property which he has made available for the use of his people (Lev. 25:23). It is not to be exploited for the enrichment of some to the detriment of others. So for example, if need compels a person to sell his land, he cannot deed it away in perpetuity, because it is not his to sell. He can only sell the number of crops to be harvested up to the next jubilee year. In essence, the same reasoning is used to explain the prohibition of slavery: the Israelites are God's servants and thus should not be enslaved to any other master. Jubilee, then, expresses a constructive social purpose, rooted in the religious conviction that all wealth is God's (Rabbi Marc Boone Fitzerman, personal communication, 7 February 1992).

This chapter of Leviticus is highly theoretical, and the jubilee year was probably never practiced as such. However, the idea of jubilee is important for two reasons. One, it signals the dangers inherent in a society where unconscionable wealth and poverty coexist and where the high priority placed on profit making threatens the integrity of the global environment. Two, this ideal represents a challenge to seek out our own solutions to the perennial problems of poverty and injustice (Rabbi Marc Boone Fitzerman, personal communication, 7 February 1992).

Is the notion of jubilee so preposterous? Perhaps. But it is interesting to note that I began writing this chapter during a jubilee year (2000), declared so by my own faith tradition, Roman Catholicism. The Catholic Church, together with several other faiths, made debt forgiveness the focus of the year. An ecumenical effort was organized to encourage the governments of the Western democracies to forgive the debt that has economically and socially crippled much of the developing world. The movement has met with some success. Recently we have witnessed the dramatic emergence of protests—Seattle, Washington, Prague, Nice, Genoa—calling upon public institutions, such as the WTO, the World Bank, and the IMF, to rethink their policies. What unites these diverse protest constituencies is the call for these institutions and governments to consider not just the maximization of monetary gain by powerful transnational corporations as goals of their policy development, but also to include global quality of life issues in the calculus of profits.

To reconfigure this equation, a few things are clear: (1) we have to put sensible boundaries on the market, (2) we have to make the economy work for the majority and not simply a few, (3) and we have to figure out a way to protect ourselves and the environment from market excesses. The measure of success must be whether every worker gains a fair share of the wealth that he or she helps to produce; safe, humane, and respectful working conditions; and the guarantee of basic human rights (Sweeny 1999:2). It should not be a political act to go to the supermarket, but it has become that for me. I wander the aisles thinking to myself: "Is the land being treated in a sustainable way? Who possesses it? Is the product I am buying produced by someone who was hurt or paid a living wage?"

I have many substantial reservations about my own faith tradition, but as Sweeny (1999:3) points out, on one point Catholicism has been unwavering: that an economic system must be judged from the position of the working person. The U.S. Catholic Bishops' 1986 Pastoral Letter, *Economic Justice for All*, says: "We judge any economic system by what it does for and to ordinary people and how it permits all to participate in it. The economy should serve the people, not the other way around" (Sweeny 1999:3). The letter supports an individual's right to work and be paid a just wage and reflects the idea that work should be consistent with human dignity and contribute to the development and fulfillment of workers.

It is my conviction that family-owned and operated farms, bound together on the land and tied to past generations through inheritance and to new generations through responsibility for a landed future, provide greater possibilities for the reinforcement of this value—land as a social trust, to be used for the benefit of others and, as my informants often told me, as the locus for family, vocation, and stewardship.

A Time to Choose

And what has been the personal cost of this colonization, this transformation? Dramatic changes in work and social practices and an accompanying loss of meaning have resulted in depression, individual and collective; crises in families in the form of domestic violence, child abuse, and alcoholism; and the loss of relationships, health, status, self-esteem, vocation, and symbols.

They have also generated a deep crisis of faith for many: spiritual faith as well as faith in government and other traditional base institutions of rural life. In too many areas of our country these crises have led to the formation of armed civilian militias, whose formation, I would argue, are rooted in a search for belonging—born out of marginalization, powerlessness, and a lack of hope.

Habermas understood the increasing infiltration of the system into the lifeworld as simply the result of the forward march of modernity. Fraser (1985) and Eley (1994) contend that Habermas failed to acknowledge the critical role of power in this process. They argue that social and historical changes derive, instead, from contests for meaning and power. Since gender is one of the principal cleavages of power and one of the central organizing categories of societies, they further contend that these dynamics of change are themselves gendered and have implications for cultural and historical ideas about men and women.

I have shown how diverging ideas about farming are both informed by and inform cultural notions of gender and subjectivity. I believe that the contests for meaning and, indeed, for the future of rural America, did not and do not just affect farmers, but also the "men in the farmers." In this study I have used pride—deeply rooted in cultural ideas about men's roles, ideas about kinship, land, history, and religion—as a tool to understand individual, local, and global cultural processes and their interrelationships. I argue that global dynamics of change, including market transitions, urbanization, migration, cultural patterns of exclusion, marginalization, and poverty, are internalized by subjects. They are personally experienced by those most vulnerable to its effects—those least capable of shielding themselves from the realities of cultural change.

The man in the farmer felt deeply ambivalent. More importantly, many men felt the ground rules were rapidly being changed and that their fundamental relationship with the public sphere was being transformed. Farm couples told me that the relationship of man to the greater community was formerly based on loyalty, the honor of one's word, an unfaltering faith in the power of hard work to get ahead or make a difference (echoed in the Oklahoma state motto, *labor omnia vincit*, or "labor conquers all things"), and a faith in government to protect its people. A man would manifest these beliefs through his sense of duty to and involvement in the community. Men believed and still believe that these virtues would be rewarded; it was upon these ideals that a man's sense of self was largely based. Men who now hold these ideas may feel naïve and out of sync with the times, which is the way that some family farmers were described to me by more industrially oriented producers.

AAM members would argue that industrial values in agriculture have provided rural men with a clear road map for success: borrowing, capital investment, good management, and expansion—but no clear destination or reason for the trip. Contemplating his considerable "success" near the end of his career, an heirless Mr. Melrose said to me as we were out working cattle, "Why am I doing this? What am I supposed to do with it now?"

Faludi, pointing to the broader national scope of this discussion, makes the argument that masculinity, like femininity, has become ornamental or superficial in American culture. She states:

> Truly, men and women have arrived at their ornamental imprisonment by different routes. Women were relegated there as a sop for their exclusion from the realm of power-striving men. Men arrived there as a result of their power-striving, which led to a society drained of context, saturated with competitive individualism that has been robbed of craft or utility and ruled by commercial values that revolve around who has the most, the biggest, the fastest.
>
> (Faludi 1999:599)

Again, if we follow Foucault's logic, it is much more than ornamental or performative masculinity or manhood that concerns us. Cultural discourses of gender are not performative, but rather are formative of external practices as well as instrumental in the shaping of internal psychological and emotional structures, an idea reflected in Jill Ker Conway's quote that opened chapter 2. There she stated, "Social systems must operate to structure the psyches of both sexes to produce their desired ideal types, and they do so by controlling what can be thought and felt" (1984:151). Despite its simplicity, her point is a significant one: emotions are political and play a role in both cultural continuity and social transformation.

Many men in my sample were clearly troubled. As the modernization of Oklahoma continued, and as the area experienced the increasing hegemony of industrial agriculture, cultural discourses that were central to the construction of men's traditional subjectivities—notions of kinship and their relation to land, ideas about gender roles, the history of northwestern Oklahoma, a sense of obligation to ancestors, their goal of family continuity on the land, and religious beliefs that position families as the natural, ordained locus of farming—were no longer supported and the new evaluative criteria did not favor men who held onto them. In other words, the cultural content of pride was being stripped away—no longer compatible with "progressive" agriculture—and, as a result, men experienced their cultural and emotional devaluation. Keeping the suicide rate of farm men in mind, some men may have also internalized a powerful message as a result of this cultural change, a message that is perhaps the ultimate subjugation: you are not good enough to live.

But more often, men responded to this cultural change with silence. Throughout this research I have struggled to understand this silence surrounding men's vulnerability. I initially chalked it up to socialization—that many men never learned or were encouraged to have a language of emotions—and I still think that this is part of the reason for their silence. But it does not explain all of it. Men in my sample were, after all, able and willing to talk to me about many moving and intimate details of their lives.

The work of Raymond Williams has helped me to recognize that perhaps another dynamic was at work. Williams (1977:128–135) calls this lack of language, this inarticulateness, the "structure of feeling." This notion describes the encounter between a new cultural grouping and the existing social order, and thus structure of feeling refers to social and not just individual feeling. Structure of feeling refers to a zone of mediation and reflection, where constant comparisons are made between what is felt or experienced and consciousness and language. The zone is always emergent and incomplete and, as Middleton (1992:206) states, it is a zone of "unfinished social relations not yet capable of reflexive self-comprehension." It is "a structural formation at the very edge of semantic availability" (Williams 1977:134), because of its strong distinction from the "official or received thought of the time." The zone is comprised of "what is not fully articulated, all that comes through as disturbance, tension, blockage, emotional trouble" (Williams 1979:168).

But some men are able to struggle past this blockage and silence, past debilitation and the cultural factors that aid their isolation, to find their voice. I contend that the American Agriculture Movement has been a powerful resource for men because it has helped them find the language that enabled them to speak of loss and vulnerability. They told their stories to one another, and in doing so, members could see the similarities of their experiences. This sharing of stories enabled them to see past the sanitized, biomedical interpretation of their experiences as "farm stress," to social and cultural factors conditioning their experiences of devaluation. Once this was done, they could begin to create a coherent response to the hegemonic discourses of industrial agriculture. They also told their stories to policy makers and the media in the service of rural advocacy and change, to preserve a way of life they believed was rapidly disappearing. Their stories of the debilitation and deaths of farmers also helped to "open up a space on the side of the road," challenging and serving as a form of "back talk" to the mythic narratives of American progress (Stewart 1996). The telling of their stories was not only therapeutic but also empowering and constructive of community.

As Fricker (1991:18) enjoins us:

> Emotions and their interpretations certainly are conditioned to some extent, but they are not wholly determined as long as we do not fail to listen to each other's stories and question the suitability of the publicly available modes of interpretation. Only then may emotions become a political force for changing how we interpret the world. If we achieve this we can assert that our emotions—if we listen to them—are not only an expression of the world, but also active participants in how the world is shaped.

Of course, this call to merge the emotional with the social is not new. Women's consciousness-raising groups of the early feminist movement skillfully brought to bear women's personal experiences in the service of individual enlightenment and to spur political change. Emotions were valued—not merely as psychological phenomena—but as embodied thoughts and as individual and social expressions of experience. Emotions were valued because their power, their "force" (R. Rosaldo 1989), to compel individual and social action was recognized. The result was that the feminist movement was able to foment a revolution in American thinking regarding cultural ideas about women and gender.

It is my hope that we can move to another such revolution in thinking regarding what farmers termed parity. "Parity is equality, you know," Mr. Gains told me. But I fear that we in America may be lost—so completely has industrial logic and thinking commodified every aspect of our lives. Instead, my hope lies in the experiences of others around the world.

As the forces of globalization collide with peoples and cultures, my hope is that technology can be increasingly utilized as a vehicle to share personal and cross-cultural experiences of the encounter between large-scale processes and individuals. I hope, too, that the sharing of experiences that results will bring to light the contradictions and paradoxes of everyday cultural and economic life, and give us reason to pause and consider fundamental questions regarding the kind of future we want to have and the meaning of human progress. And finally, I hope that this encounter will generate the force of feeling required to move us to a different kind of future, one of greater social justice and parity—a future that includes us all.

APPENDIX

Wide, Open Spaces (1993)

THIS PIECE is the result of a period of unproductiveness during fieldwork spurred by a conversation I had with one of my informants. During an interview session with the Melroses, Mrs. Melrose asked me about myself and my own ties to Enid and northwestern Oklahoma. Her questions made me anxious and led me to begin to examine what community meant to me. This piece originally had two purposes: to formulate a response to my informants regarding my own sense of community but also to *desahogarme*, a Spanish word meaning "to relieve one's self through the expression of feelings" or, literally, "to un-drown oneself."

With respect to theoretical issues regarding emotions, this essay may serve an additional function: to demonstrate one potential vehicle for the reconciliation of the individual and the cultural. *Out of Place*, the memoir of

Edward Said (1999), serves as an example of this reconciliation. Through an examination of his youth as part of the Palestinian diaspora, Said elucidates the development of a system of "personal meaning" (Chodorow 1999) and "emotional force" (R. Rosaldo 1989) that was to later fuel his activism for the Palestinian cause.

The writing of this piece helped me to arrive at some conclusions about community and my own motivations and force to conduct research on gender and emotions. I hope it will help to illuminate aspects of Oklahoma culture or at least give a sense of a system of personal meaning that a researcher brings to his or her project.

NO HAS MUERTO. HAS VUELTO A MI

No has muerto. Has vuelto a mi. Lo que en la tierra
—donde una parte de tu ser reposa—
sepultaron los hombres, no te encierra;
porque yo soy tu verdadera fosa.

Dentro de esta inquietud del alma ansiosa
que me diste al nacer, sigues en guerra
contra la insaciedad que nos acosa
y que, desde la cuna, nos destierra.

Vives en lo que pienso, en lo que digo,
y con vida tan honda que no hay centro,
hora y lugar en que no estés conmigo;

pues te clavó la muerte tan adentro
del corazón filial con que te abrigo
que, mientras más me busco, más te encuentro.

YOU HAVE NOT DIED. YOU HAVE RETURNED TO ME

You have not died.   You have returned to me.
Some part of you is at rest underground,
buried by men, but earth does not hold you,
because I am your true resting place.

Inside my chaotic, anxious soul
that is your legacy, you continue
to fight the longing that accosts us
and, since the crib, banishes us.

You live in what I think, in what I say,
a life so profound there is no center, no time
or place you are not with me;

death has nailed you so deep inside
my filial heart, now your sanctuary,
that the more I search myself out, the more I find you.

—Jaime Torres Bodet (1902), translation by Claire Moor Aaronson

The Filial Heart

In my mind, the details of my father's life in Cuba are sketchy. I know that he was born and raised in the country, the son of poor peasants. He went to Havana and became a police officer. Somewhere along the way he got involved in politics. He was a political prisoner under Batista for three years in *La Cabaña*. He initially supported the Castro revolution and when the revolution was victorious, he was given a position in the newly formed government. Remember Castro's historic visit to the United States that included a stop in Harlem? My father was there with him. But as my mother tells the story, my father became disenchanted with Castro when he forged closer ties with the Soviet Union. After my parents married, they left Cuba, never to return. It seems to me now that leaving Cuba was not an isolated event for my father, but a process that defined his life and, consequently, our lives.

My parents' early experience in the United States is clearer to me. They moved to New York City in 1961 and settled in the Washington Heights neighborhood on 172nd Street by the Hudson River. My brother Nelson was born in 1962 and I in 1963. To support us, my father worked as a waiter at the Chateau Madrid Restaurant and Club and later took up diamond dealing. In many ways he was the American ideal of what an immigrant should be: an aspiring, hard worker who wanted a better life for his children and who readily took up the culture of his new home.

My father was bright, energetic, and an eternal optimist. He was funny: when speaking with *Americanos* who had trouble understanding his heavily accented speech, he would say, "Whatsa matta? You don't speak-ee English?" And he was spontaneous. For example, we might all be sitting around the living room involved in our own pursuits, when my father would get up, walk over to the record player, put on some meringue beat music, begin to sway the hips of his short, heavyset frame, smile brightly and say, "Let's dance!" My brother and I would jump up, laughing convulsively. My mother would sigh, roll her eyes, and say to herself, "*Ay Dios mio.*" After some coaxing, my father would take her hands, lift her off the sofa and they would be off, expertly gliding across the living room floor—my brother and I trying to imitate the best we could.

He was a man of many gifts, gifts that he was always willing to share. When I was young, my father would give me a quarter for every story that I wrote. I would leave my white, single, looseleaf sheet of paper at his place on the kitchen table for him to read in the middle of the night when he arrived home from his restaurant job. Though I never witnessed it, I imagine he sat there, cigarette in hand, with a closemouthed grin on his face, reading and wondering about the future of his children.

In the morning, I would wake up, jump out of bed, and scurry in my socks into the kitchen. There it would be: the shiny coin that represented an ensured supply of Hershey bars.

When I would arrive home from school, my father and I would talk about what I had written. He would tell me who his favorite character was, comment on an observation that I had made, or ask me why I had chosen a particular setting. I ask myself now, "How did he know? Where did this man of poor peasant origins learn to think beyond his world that he was able to teach us not to have small minds, to be ambitious and to have—his special gift to me—a passion for the written word?"

He was an opinionated man and was not bashful about sharing his views with anyone who would listen. He was particularly passionate about politics and was a voracious reader of news and political journals and commentary. Ever since I can remember my brother and I were surrounded by political discussions. The political talk was often focused on Cuba and its future and always dealt with the evils of communism and its workings in the world. There was a real vehemence or aggression in his beliefs about communism. The vehemence was more than was warranted by the simple rejection of an idea by an individual. His lived experience provided the emotion. But it was his personal experiences about which we never heard. Were there regrets about leaving? Was it difficult to lose your homeland? Was it wrenching to be away from your family? I think it was easier for him to focus on communism; it was more tangible and ultimately less threatening.

For my father, communism was a basic orienting principle, a tool that he used to make sense of the world. Its influence, he perceived, was not limited to politics, but pervasive in all aspects of life. His thought processes were a structural analyst's paradise—communism was associated with everything in the world that he felt was wrong. Communists were trying to cause trouble in Latin America and Africa. They were responsible for the drug problem in our cities. They were the social agitators and demonstrators compromising America's morals. But my father associated communism, too, with things I clearly did not: the African American civil rights movement, the women's movement, and the Vietnam antiwar effort.

He knew that I thought about things differently than he did. In fact, I think that my father felt that, despite his best efforts, the demon seed of liberalism had been planted inside of me and he saw it as his job to root it out as quickly as possible. When I presented other visions of the world that did not conform to his, he would come unglued. My father was a controlling man. He was also a nervous man.

One day, in my early college career, I was reading, sitting in my bedroom, in my favorite chair, by my favorite window looking west. The sun was setting, and the light, in combination with the clouds, was producing dazzling hues of red. My dog was lying at my feet, keeping them warm. "This must be what goes on in heaven," I remember thinking, when the door of my bedroom was thrown open. My father charged toward me, arms waving in the air, yelling in a loud, angry voice. The word that I could make out in his initial verbal assault was "*Comunista!*"

Communist.

I had been careless. I knew better. I had left a book on Karl Marx lying on the kitchen table. He had the book in his hands. I had broken the number one rule for children of Cubans living in exile: if you have leftist leanings, hide them away so that no one will ever see. Those thoughts are dirty, immoral, and can give your family a bad name. Most importantly, those thoughts are a betrayal to the people who love you the most.

"Why are you reading *esta mierda*?" "You're becoming a communist, aren't you?" "This is what you are learning at University?" "Didn't we teach you anything?" "Why do you think we left Cuba?" and with that last word—Cuba—the book left his hands; he threw it with such force and passion that one would have thought that he had Fidel himself by the ankles. Had it been *El Comandante*, he would have not survived the impact with my bookcase.

I, forestalling a similar fate for myself (my dog was long gone), pleaded innocent, "Papi, it's a book that was assigned to me in my anthropology class." After some elaboration and negotiation, this tactic worked, but it prompted an indictment of the American university system which, according to him, was being overrun by Russian, communist infiltrators.

My father began to walk out of my room, satisfied. Although I escaped censure on a technicality, he knew that he had made his point. He walked out of my room, turned the corner and burst into my brother's room. Instead of verbally assaulting my brother for reading "party" literature, he accused him of never picking up a book—of not reading anything at all.

When I remember that scene, it's almost funny. But I don't kid myself; this scene and several others like them had their effect. My father hated communists. My father hated what Cuba had become. He tried to pass on those feelings to us. For me, hating was too active. His aggression about Cuba produced a silence and an empty space within me instead: the nation of Cuba slipped from my mind; it simply ceased to exist.

When I was eleven, we moved away from New York City. My father, concerned about the increasingly visible signs of urban decay, decided that he wanted my brother and me to have the benefit of a country upbringing. We had ties to the Midwest through my father's best friend from childhood who was a physician in Arkansas City, Kansas. Together they bought a large motel in Enid, which my parents managed.

For my brother and me, the move was a major adjustment. We had to become accustomed to living in two worlds: the "American" culture of school and friends and the Cuban culture of home. With time, however, we became skilled cultural translators for our parents, and we learned to selectively negotiate pathways that led to our relative success in both worlds. My mother was ambivalent about the move to Oklahoma from the start; her preference was clearly for Miami. After the move I think she felt that she had been set adrift. An avowed city person (she had spent her whole life in Havana and New York), she suddenly found herself amidst wheat fields and cows. Frustrating the situation further, there were few individuals with whom she could share her experiences; she lacked proficiency in English and, at that time, 1975, in Enid, there were very few Latino families, regardless of national origin.

My father, on the other hand, fit right in from the start. He joined civic groups, the Chamber of Commerce, and became an avid supporter of the local Air Force base. He made friends quickly and easily. My father embraced Middle America and Middle America welcomed this outgoing, energetic, Republican Cuban with open arms. It was a natural fit.

My father wanted us to be as enthusiastic about our new lives as he was. I remember asking my father something about Cuban culture one day. He waved away my question with his hand, "Forget that Cuban stuff. You're an American and you need to know American ways to be successful." And with time my brother and I did just that: we began to forget.

One day, years later, my brother and I were sitting around the motel lobby, bored. We were flipping the channels on the television. We stopped for a moment when we heard Spanish on a station; at that time, you couldn't get any Spanish programming in Oklahoma. It was a comedy show and we quickly realized that it was about a Cuban family in Miami. The show was called "*Que Pasa, USA?*" My brother ran over to the telephone switchboard to call my parents at home to encourage them to turn on the television.

The program immediately drew us in—it so closely mirrored our own family: there were the children trying to do the best job possible of living in two worlds; there was the father, the ultimate arbiter of cultural meaning; and there was the mother, the reasonable peacekeeper. When my brother and I arrived home after the show, we were all buzzing with energy and excitement. We made joking accusations: "You are just like that!" or "You say that all the time!" We were so happy. We had begun to think that there was no one else on earth like us.

Dependency Theory

My brother would get upset with me when I would cheer for women on TV game shows.

"Why aren't you ever for the guys, Eric? God, you're weird!"

I think my mother had something to do with that. I didn't feel particularly comfortable playing with the boys on our street: I was fat, I wasn't fast, I was clumsy, and—sin of all sins—I was caught playing dolls with my female cousins one day. This sealed my fate with the neighborhood boys. So I became my mother's sidekick.

Our days were spent running errands, shopping, visiting relatives and friends, or at home: she doing endless housework as I read, wrote, listened to music, and played. We were good companions; we enjoyed talking and we have always had the gift of making each other laugh.

Through time, I settled in under her. It is difficult for me, even now, to separate anything within myself that did not have its origins in her. I remember sitting quietly by as my mother spoke with her friends. I followed these conversations. Through them I learned not only what to be, but how to be it. The characters involved in my mother's stories became important to me and I quickly internalized the implicit values of the stories. From my mother I learned the intent of a half-raised eyebrow or a tilted head. I learned to nod and punctuate my speech with "*si*," "*seguro*," and "*claro*" in affirmation of messages received during conversations. I learned what to feel, when, and how. From my mother I learned to give love freely and laugh heartily and with honesty. Because of her and her friends, I have always known that women are integral and complete without men.

But it seemed that my mother had a degree of wholeness away from my father that, with him, was not allowed to be. Occasionally he and my mother would have a disagreement. My mother would say something; he would interrupt her. I remember thinking, "*Trata otra vez, Mami, tienes razón!*" (Try again, Mom, you're right!)

My mother would take a different tack, she would say something else, with increasing volume. He would shout.

"*Haga que te oiga, Mami!*" (Make him hear you, Mom!)

She would shout. He would shout louder.

"*Mami, no te pares!*" (Mom, don't stop!)

She would say something under her breath. He would mock her.

"*Está bien Mami, mejor que no digas más nada. No vale la pena.*" (It's OK Mom, it's better that you don't say anything else. It's not worth it.)

I learned resistance from her too, incorporating her style without much consideration of its effectiveness. It was clear that in our family, our voices were not always heard. We were relegated to creating mental space in the places where my father did not predominate. My mother and I formed an alliance that, at the very least, ensured that we always heard each other. We developed our own language and values. The alliance was rich and psychologically safe.

My father and brother were jealous of this, I think. We had codes that they couldn't decipher. If my mother and I were of the same mind on a certain issue, my brother and father knew that they faced an uphill battle. It is hard to put our secret into words, but I know that it had something to do with sensitivity to others, emotional talk as legitimate talk, social perceptiveness, and our perceived role similarities. She cheered my victories and I hers.

I wanted to go away for college, but my father wouldn't let me. He said that I should go to the local Christian university for a couple of years. Then, if I still wanted to go back East, he would consider it. That was not his intention, however. He thought that I would become webbed in friendships, activities, and interests and ultimately forget my plan. I did not.

I don't know why it was so important for me to go away to school. Partly, I was convinced that there were wonderful things to see and experience elsewhere and I felt that they were passing me by. I know now that this was not a sufficient reason to warrant the depth of conviction I felt about leaving. In thinking about my past, I now see that I wanted to escape the web that held me in time and space and didn't let me be anything else.

After two years at the local university, I began to apply to East Coast colleges in secret. Only my mother knew. She provided me with money for the requisite application and transcript fees. One day my father found my completed application to Bennington College in the outgoing mail. He removed the application from the pile and kept it until we were all three in the kitchen.

"What are you doing?" he asked. His question was without context but my mother and I knew exactly what he meant.

Silence.

He pulled from his pocket the envelope that contained the application. It had been folded and refolded in a myriad of ways. My heart sank.

"What is this?" he asked.

"*No es nada*, Papi. I just want to see if I can get accepted."

"*Por qúe estás gastando tu tiempo?*" (Why are you wasting your time?)

I sat down, spirit falling, a lump rising in my throat. I said nothing.

"If you go to Vermont, don't ever bother coming home," he said. This dropped like a bomb. Groping for orientation, I couldn't respond.

He then turned to my mother; this was his usual strategy of divide and conquer. "Why are you encouraging him?"

"I'm not encouraging him. I want him to do what makes him happy."

I grew up hearing—every day—that everything my parents did, they did for my brother and me. My happiness was an arguable point. By this time I had steadied myself.

"Papi, I want to go," I mustered. I looked at him seriously. This was his cue. We had an unspoken agreement that if I felt I really needed something, I would circumvent the usual "Papi, please" begging routine, and ask for what I wanted directly. He, in turn, would get to the heart of the matter and address me with concrete reasoning.

He looked at me with concern, "*Hijo*, you're not ready to leave home. You are too naïve; you trust too easily. People will take advantage of you. If anyone is ready to go, it is your brother. He has had more experience. He has earned money. He can take care of himself. You are not ready. I also don't think you can handle any more academic pressure than you have here."

I felt deeply betrayed and I was angry. I had worked hard to please my parents; I was, after all, what they wanted me to be. I grabbed the crumpled Bennington application that my father had by now laid on the kitchen table, walked out of the house, drove to the community post office, and mailed it.

Around the time that admissions decision letters were to be mailed, I became riveted to our mailbox. I began a pattern of going to the mailbox daily, opening it carefully, as if my carelessness would make the letter I had waited for for so long disappear. One day, I peered inside and the letter from the admissions office was sitting on top of a huge pile of mail. I gathered all the mail in my left arm and held the letter from the college separately in my right hand. I walked slowly, ceremoniously even, towards the house, again as if my misstep would jumble the words and change the intent of the letter.

My father was sitting at the dining room table reading the morning paper when I came in. My mother was dusting in the same room. I placed the mail on the table and my father began to examine it. I walked behind him and waved the letter overhead until I caught my mother's attention. I began to walk back to my brother's bedroom and I heard my mother start in behind me. We went in and I closed the door. We sat on my brother's bed, looked at each other and took a deep breath. She put her hand on my arm as I opened the letter and began to read. I read slowly, not comprehending the words so neatly typed before me. The last sentence of the first paragraph was clear, "We are pleased to offer you admission to the Class of 1985."

"Mami, *me escogieron*," I said slowly. "*Me escogieron!*" We leapt to our feet and jumped up and down holding on to each other.

I'm not sure what made my father change his mind and let me go. Partly, I know, it was because of my mother and partly, I think now, was because my father knew how much it meant to me. He let me go because he loved me that much.

I was terrified on the day I was to leave, 10 September 1983. It was 5:30 in the morning. My friend Mary Porter was waiting in her car to take me to the Oklahoma City airport, one hundred miles away. My mother sensed my terror. She pulled me close and told me not to be frightened. I began to cry. She told me that I had learned a different way—that I could get along with anyone in the world—and that that was my strength and a gift. She told me I would do fine.

She kissed me and we hugged. I realized then that much of whatever strength that I have comes from her. She created a rich life for herself and for us. She showed me that I could do the same. I cheer for women on game shows because in doing so I cheer for people like my mother, people like me.

My father walked me to the car. He was unable to speak and began to weep openly. He never said a word as I was leaving. He put his arms around me and we hung on tight for a long time. I think we both knew it was the last time we would see each other.

Halloween 1983

I remember the hollowness of my body when my brother called to tell me our father had died suddenly of a heart attack. I remember my wonderment at how such an empty space could produce so many tears. I remember, an hour later, beginning the midnight car ride to JFK Airport in New York City to catch the earliest morning flight to Tulsa. I imagined a movie set: myself in the back of a dark carriage drawn by black horses racing through the relentless snow, their breath visible as they heaved their heavy load forward. I remember as we traveled, my new friends shifting uncomfortably in the car as I periodically became aware that my life—our lives—would never be the same. At those moments I would wail. I remember reading the Psalms on the plane and finding a comfort within them that I had never experienced before, or since. I remember thinking the plane was going to fall and that I was going to die.

I remember standing in the back of St. Francis with my mother and brother in preparation for the procession towards the altar. As we stood there, in the church in which my brother and I had grown up, the baritone bells began to sound directly overhead. They struck deep within me reverberating forcefully against the walls holding together the emptiness inside me. After the funeral in Oklahoma, I remember the two day vigil, sitting at my father's wake in Miami, meeting many people I didn't recall having met before. I remember the flowers—so many flowers—wreaths upon wreaths stacked one upon the other covering the walls, symbols of a life lived before us, apart from us, beyond us.

And I remember my mother giving my father one last kiss. She silently took two steps forward, family crowded around, and bent slowly towards the body. My favorite cousin Annie pressed my elbow, whispering quietly to herself, "*Ay Virgen! No puedo más!*" (I can't go on!) My mother gently placed her lips upon his. A seal.

They closed the casket and buried him. It was my birthday.

A Stranger on the Land

I could hear the geese before I could see them. It's winter in Oklahoma. The sky is gray and the clouds are hanging low. I'm standing in Mr. Melrose's pastureland. He is mending a fence that an errant steer has broken through. I'm close by, leaning on his red Ford pickup, asking questions with pen and pad in hand. He's discussing the American public's lack of understanding of the process involved in getting food to supermarket shelves.

I'm searching; I am intent on finding the geese. Mr. Melrose notices me and points to them.

"There," and he goes back to circling his pliers, pulling the barbed wire tighter and tighter.

In the instant my eyes meet the gaggle, I hear Mr. Melrose say that he is in financial trouble.

"I'm one of those farmers your project is about, Eric."

As he continued to talk, I listened intently though my eyes followed the gaggle across the sky until it was nearly out of sight. On this farm, only thirty miles from where I grew up, I knew not to make a spectacle of my acknowledgment of his situation, of his pain. This I had learned.

As an anthropologist, my current research focuses on Oklahoma farmers. I am interested in their ideas about masculinity and male roles and how these notions influence their responses to farm financial trouble—which in many cases has led to farmers losing land which had been in their families for generations. I was drawn to this research because it held the potential of answering, for me, central academic questions: how ideas about gender relate to other spheres of social life, how men's roles and identities are socially constructed, and how culture influences our emotions and sense of self.

What I didn't realize as I began, was that I was also drawn to this research project because it held the potential of answering personal questions.

I think that I mentally placed my father in the same category as the farm men, the subjects of my research. To me both personified masculinity, both represented aspects of the American dream, and both, in different ways, had lost their land. A part of me wonders if I, unconsciously, went to my subjects with the hope that they could tell me, offer me, something my father could not.

I sat enthralled listening to their stories of pioneer courage and perseverance: surviving tornadoes, preparing for approaching dust storms during the Dust Bowl, women working in the fields while pregnant—coming to the house just long enough to give birth, mothers and fathers taking turns staying up during the night to stand guard against encroaching prairie fires.

Their stories were familiar to me; I had grown up in Oklahoma hearing them and I realized that they had become part of my own sense of self, my own personal lore of which I was proud.

But their stories, to me, were incomplete. There was something else that I wanted them to say, which they, of course, could not. That something had to do with my other history and my other community. Where were the other stories? I began to wonder. Where were the stories of revolutions, those of divided families, of emigration to a foreign land, of immigrant challenges and courage? My research, unexpectedly, sparked my interest in my ethnicity, in my identity as a Cuban.

As I became aware of my own Cuban-ness, I realized that designating myself as Cuban felt a bit premature, like the insecurity that one feels when a decision is made without much information. My brother and I share a conception of being Cuban that resembles anthropologists' ideas about culture at the turn of the century—that our cultural heritage is composed of lists of artifacts or traits. Because of our very limited exposure to other Cubans, my brother and I have little in the way of stories or other contexts that draw meanings and connections between traits. The only connections, that we are conscious of, of course, are our parents.

My father was a mentally imposing man. He was like a compass or dominant landmark that I could always use to plot my course. Sometimes for us it was difficult to see around him and get a clear picture of things. His opinions so strongly influenced my own that while he was alive I didn't know what to question or, I suspect, even how to question. While he was alive it was difficult to examine my identity as a Cuban because he so closely aligned not only the island, but I think Cubans generally, regardless of their location, with communism. He felt comfortable in America's heartland because he wasn't reminded of the past, of a great disillusionment, a great disappointment, and a great pain. Thus my father's movement away from Cuba during his life made my link to the past and to fuller self-understanding very tenuous. When my father "left" Cuba he did so as well for the rest of us.

My father's death was a stunning loss that left me unmoored for a long time. I think it was unfortunate that my leaving home and his death happened so closely in time. Leaving home for college was a way for me to leave his domination; it was a form of rebellion that did not directly challenge his authority. It was a process that I began somewhat consciously, but his death took from me the element of choice. I have had to continue the process of "leaving" and "becoming" without the benefit of a dialogue with him.

So my only link to the past now is my mother. As my father's death moves farther into the distance, it is easier to question the past and its implications for the present, its implications for me.

My mother often sits in my brother's room by the window with the sewing in her lap. She is relaxed when she sews and we often have our most meaningful conversations then. I scour my mother's memories for ideas, details, and perspectives. I usually begin by saying, "Mami, tell me about . . ." In response, she will set the sewing on her lap, sit back in her chair, lift her head to meet my gaze and begin. She is a wonderful storyteller. Her face is alive with expressions: her eyes open wide, her eyebrows move up and down, she gesticulates animatedly, and her demeanor is open and inviting. Her stories of Cuba are not about grand political ideals, but about people. They are filled with personalities, emotions, and the dilemmas of familial relationships. In essence they are about human connection. These are the stories that she taught me to understand.

After a particularly dramatic story about our family, I said, "Why didn't you ever tell me that before?" She answered me, "You never asked. There is a lot you don't know." The problem in being the only one of a kind in a place is that you begin to forget what it is that makes you different. I suspect I never asked because I never learned what to ask.

All our talk has made me very curious about the Cuba of the past and the Cuba of the present. We have even started to make plans to visit the island. I want to learn more and part of it is very personal. I want to know how I fit in terms of what has come before. I want to know that I'm not a statistical, demographic aberration, but that my life has and will have meaning in its own terms. I want to know that my story—perhaps this story—will make sense to someone, somewhere. That he or she can read it and say, "I am like him and he is one of us."

Mr. Melrose and I are riding around in the pickup feeding his cattle, which graze on several different farms. I have learned that these are the moments when he is most comfortable talking; our conversations are mostly spontaneous and, much to his delight, a tape recorder is nowhere to be seen. Today I am asking him why Oklahoma farm men have such a hard time deal-ing with financial crises and why it is so difficult for them to reach out for help. He says he's not sure.

But then he adds, "We sure are a proud people, Eric, aren't we?"

This question is an offer of acceptance, an invitation to belong. My ex-perience as a Cuban-Oklahoman, as my father's son, as someone who has always felt a part and yet apart, has taught me to accept these invitations whenever and from whomever they come.

I smile and answer, "Yes, we are."

NOTES

Introduction. Homework

1. To honor my promise of anonymity to the people with whom I worked, I used pseudonyms for both individuals and communities in this study. I made an occasional exception for professional and personal contacts who did not mind having their identities revealed. To further ensure the anonymity of my informants, I also changed, in many cases, irrelevant (to this study) details of their lives which could serve as potential identifiers. For instance, I often altered the direction I traveled out of Enid to arrive at an informant's homeplace and, on occasion, the number, age, and sex of participants' children, or even the occupation of a spouse. The importance of these measures became evident to me during the course of a conversation with a colleague, who along with me was conducting "first responder" trainings in rural communities to help volunteers recognize and respond to farmers in

crisis. During this conversation, my colleague (who was a farm woman) asked me on whose farm I was working as part of my research. I told her that I couldn't reveal that information, but that the farm was located about thirty miles west of Enid. She then asked me how large the farm was and how many head of cattle were on the property. I answered her questions and we continued our conversation, leaving behind the topic of my research. About ten minutes later, she suddenly said, "You're working on the Melrose place, aren't you?" I was amazed at her correct guess and asked her how she had arrived at that conclusion. She responded that she mentally began to drive up and down the roads thirty miles west of Enid until she came upon the Melrose farm (because of the grid pattern in which western Oklahoma is laid out, most residents have a keen sense of direction and distance). She realized that that operation closely matched the acreage and number of cattle I had mentioned. Fortunately, my colleague is a psychologist and understood the importance of confidentiality and anonymity. I think this conversation speaks to the very public nature of farming, a topic that will be explored fully later in the book.

2. This figure is from the joint study conducted by the *Tulsa World* and Oklahoma State University which was featured on the front page of the *Tulsa World* in an article by Mark Lee entitled, "Study Shows Suicide High Among Farmers."

3. There is already a significant body of work that challenges the common understanding of emotions as biological and psychological phenomena. Much of this work acknowledges that individuals draw from the discourses available to them in constructing their subjectivity—their sense of who they are. These scholars argue that language is critical in this process of construction. It is through language that "actual and possible forms of social organization and their likely social and political consequences are defined and contested. Yet, it is also the place where our sense of self, our subjectivity, is *constructed*" (Weedon 1987:21). Included in this subjectivity is the individual's emotional world. I contend that to understand social action as a product of identity, we must understand emotions as "cognitions—or more aptly, perhaps, interpretations—always culturally informed" and mediated through social practices by the stories that individuals enact and tell (Rosaldo 1984:141–143). There have been numerous studies about emotions, including works by Abu-Lughod (1986), Lutz (1988), Reddy (1997), M. Rosaldo (1984), R. Rosaldo (1989), Scheper-Hughes (1992), and Yanagisako (2002). Few studies, however, have examined the intersection of gender and emotion. Of these, three examples are the works of Abu-Lughod (1986), Scheper-Hughes (1992), and Yanagisako (2002). These studies demonstrate the utility of examining emotions from the standpoint of actors in particular positions of power within a social structure. Furthermore, they suggest a link between emotions as gendered experiences and beliefs and actions.

4. Visweswaran has called for anthropologists to do their *homework*: to return to, and study, their "own neighborhoods and growing-up places" (1994:104). She sees this as an important step in the continuing process of the decolonization of anthro-

pology. By calling into question the epistemological core of anthropology—framed by the departure and return of fieldwork abroad—Visweswaran highlights the problematic of the "ventriloquist fantasy," of speaking from a place one is not from. It seems to me that her call for homework also raises some important issues regarding the roles and responsibilities of anthropologists in their home communities: What responsibility does the citizen/anthropologist have to better her community, to intervene? Why do we do anthropology at all? I would add that Visweswaran's (1994:113) felicitous phrase—"Home once interrogated is a place we have never before been"—is true not only because anthropologists, already "positioned" in their communities, would begin to look at their growing-up places in new ways when embarking on research, but also, perhaps, because their research may help to bring about social change.

1. The Invitation to Die

1. *Country* (starring Jessica Lange and Sam Shepard) and *The River* (with Sissy Spacek and Mel Gibson) were notable films that were released in the mid-1980s dramatizing the plight of family farmers. Dudley (2000) notes that Lange, Spacek, and Jane Fonda (*The Dollmaker*) capitalized on their celebrity status to testify before select committees of Congress to bring increased attention to the experience of struggling farm families and urge agricultural public policy changes.

2. In this study, I use the word "crisis" to describe the experience of my informants, since this is the term they consistently employed to talk about their own experiences and feelings, even many years after their initial episode of financial and emotional distress. It was one strategy the participants of this study used to communicate to me the chronicity of the 1980s crisis and the nature of the changes they continue to witness in their communities.

3. Though many medical anthropologists have implicitly and explicitly addressed the relationship between health and inequality in their work, the issue is only beginning to reach public health circles. Especially influential in that context is the work of Paul Farmer (1999, 2003).

4. I conducted research in the following counties: Kay, Noble, Ellis, Garfield, Grant, Major, Woodward, Kingfisher. All were part of the Cherokee Strip (or Cherokee Outlet; though the "Strip" and "Outlet" are geographically distinct, these names are used interchangeably today), except for Kingfisher and Ellis Counties, which are contiguous to the Strip.

5. The work of Barlett and Conger (n.d.) demonstrates that the sustainable agriculture movement has generated a third "vision of masculine success" that is less competitive and individualistic and more sensitive to global environmental concerns.

6. I understand discourses to be constituted by language and practices that organize social life and social reproduction. The social world that is created or persists allows individuals to understand and to determine their internal as well as ex-

ternal sense of belonging and to create the individual and social capacity to recognize what is foreign or "other" (Terdiman 1985:54).

7. Dudley (2003:178–179) does acknowledge the importance of the strategy of family continuity, but ultimately does not see it as being incompatible with the interests of individual achievement and profit maximization.

8. I utilized the research model suggested by Yanagisako and Collier (1987:38–48) to understand gender within particular cultural contexts. Briefly, the model is composed of three principal steps. The first step is to conduct a cultural analysis of meaning. This entails coming to an understanding of the "socially meaningful categories people employ and encounter in specific social contexts and what symbols and meaning underlie them" (1987:43). The second step involves the identification of the metaphors of inequality. Here, I replace Yanagisako and Collier's focus on "systemic models of inequality" (1987:42) with metaphors. Models, I assert, are primarily useful because they allow an economy of description while making sense of diverse cultural dimensions, social organizations, and interconnections. As Yanagisako and Collier state, "All attempts to understand other cultures are, by their nature, comparative" (1987:43) and, I would argue, predictive (see Arensberg 1955:1145–1146). My concern is that a focus on models may obstruct the interplay of social forces and diverging goals of individuals, and perhaps predispose us to preconceived frames of reference. I prefer metaphors: an identification and focus on the manner and the domains in which ideas about inequality and gender are discussed, negotiated, and acted upon. But ultimately, models and metaphors share the same goal: "to understand how inequality is organized in any particular society" (Yanagisako and Collier 1987:42). The final step in understanding gender in a particular cultural setting is through historical analysis. Historical analysis is useful because over time it demonstrates the consequences of the interactions between individuals (and their actions) and social structures. This historical frame allows a comparative focus within a particularist approach (such as this study) by demonstrating that "things haven't always been this way" and thus elucidating the cultural basis of naturalized discourses. This approach also "broadens the temporal range of our analysis of social wholes by asking how their connection with the past constrains and shapes their dynamics in the present" (1987:46).

2. The Nelsons

1. The tractorcade was a form of protest or demonstration employed by organizations that supported family farmers. Typically convoys of tractors would travel to a designated location such as a county seat, town center, or even Washington, D.C. Their impressive presence and the occasional inconveniences they would cause were intended to call attention to the plight of the family farmer. The tractorcades were also intended to raise the consciousness of urbanites regarding the connections between city and rural economies and issues of food security and accessibility.

2. Kimmel (1996:26) calls this same-sex social process "homosociality."

3. Jane Adams, personal communication (July 2, 2002).

3. Creating Oklahoma

1. It is important to note, however, that the video, "Spirit of '93: Stories of the Cherokee Strip Land Run," produced by the Cherokee Strip Centennial Foundation (and written by Pat Bellmon) to commemorate the one hundredth anniversary of this historic event, told the story of the Cherokees' forced relinquishment of their western lands in Oklahoma. The video makes it clear that this result was obtained through less than ethical means by the federal government.

2. Renowned Oklahoma historian Angie Debo (1940) described the various laws enacted by Georgia, Mississippi, and other states that effectively tied the hands of tribal members to legally contest the encroachment of Indian lands and infringement of tribal rights. Perhaps more importantly, she points out that the clear intent of these legislative acts was to hobble the autonomy of the tribes. For instance, in 1830 Mississippi enacted legislation that made Choctaw and Chickasaws citizens of that state and barred members of these groups from holding any tribal office. Georgia passed a law that prohibited the Cherokee legislative body to meet, except for the explicit "purpose of ratifying land cessions" (Debo 1940:4). In addition, that same state encouraged—with impunity—the continued plundering of Cherokee lands by passing a law that forbade any tribal member from bringing suit against a white man (Debo 1940:4).

3. Dudley (2000:140) also claims that farm loss is experienced as death: "a socially constructed kind of death, one that comes from having 'everything you have' forcibly taken away." In some ways, I believe she and I are both speaking about the "death" of the farmer's identity and, perhaps, that of the farm family as well.

4. It is interesting to note that particular farms often are used as landmarks. For instance, in giving directions, an individual might say, "Head south until you get to the Smith place, take a left to the Nowak place, and then head south again for a mile." I wondered whether the desire of families for land to remain in the family name was related to this public acknowledgement—that the "Smith place" would always be known as the "Smith place." However, I could not arrive at a consensus on this issue with my informants. Some agreed that there is a certain sense of prestige and immortality associated with this practice that families might be interested in perpetuating. Others disagreed and did not believe this was a motivating factor for the desire of land to remain in the family name. As evidence they cited the fact that a location may continue to be known as the "Smith place" even though the Smiths had not owned that property for over a generation. Thus ownership did not always correlate with place name.

5. For an excellent description of the principal features of industrial agriculture, see Barlett (1989).

6. These values have become so institutionalized that they almost seem self-

evident, or natural. But as Fitzgerald (2003) documents, the emergence of the industrialization of agriculture can be discerned in the early twentieth century and is the result of specific historical and cultural factors that coalesced during that period. She argues that the principal metaphor driving this practical, ideological, and cultural change was that of the "farm as factory." Inspired by Frederick Taylor, the father of management science, the professionalization of a new breed of agricultural specialists was propelled by a period of depression in the agricultural sector in the 1920s, not too unlike that experienced in the 1980s. These new agricultural economists, engineers, and managers believed that the standardization and routinization of tasks and processes, which had been successfully adopted in factories, could serve to increase the efficiency (itself a value gaining prominence at the time) of agricultural production. Their principal impulse was toward quantification and mechanization. Through their work and emergent institutions, American farmers—farmer by farmer—became convinced of the need to change the ways they produced food. But the process was not completed at one time, and industrial values are still in competition with other agricultural goals and strategies, as this study suggests.

7. Barnett (2003:170–171) discusses a number of interrelated reasons why agricultural economists seemed to be caught off guard by the farm financial crisis of the early 1980s. First, they failed to recognize that contractionary monetary policies would be the only politically acceptable strategy to deal with the inflationary economy. Second, he points to agricultural economists' practice of excluding macroeconomic factors as exogenous variables in their models. Third and more generally, Barnett asserts that economists have become accustomed to excluding politics and history from their analyses because these are difficult to quantify. The result is that agricultural economists failed to appreciate the historical and political constraints under which the makers of public policy functioned.

4. The Good Farmer

1. Postmodernism has not been without its critics; for an excellent critique of postmodernism and relativism see Bowlin and Stromberg (1997).

2. Most of the family-oriented farm couples I interviewed were members of the American Agriculture Movement, an organization that explicitly promotes family-centered values in agriculture. Because of this, I believe, the clarity of their vision regarding the role of families on the farm was more readily discernable than more industrially oriented families.

3. It should be noted that this perspective is not universally shared. In her historical examination of the role of women in rural Nebraska, Fink (1992) makes a compelling case for questioning the assumptions and romanticization of agrarianism, which has traditionally been understood as being supportive of women's particular capacities, power, and contributions to the farm. Instead, Fink suggests, agrarianism has functioned as a myth validating "the dream of the individual

farmer working his own land and reaping his own profits" (1992:191) while deemphasizing the labor contributions of women and other family members. Further, Fink asserts that women's power was ultimately founded on the nuclear family, "which constrained any power they might have garnered through their economic activities" (1992:190). Women were not shielded from the gender oppression their urban sisters experienced. But unlike them, rural farm women found themselves highly isolated—emotionally and physically—and consequently denied the benefits of the "social protection against violence and exploitation" afforded to women in urban areas who were able to support and aid one another.

5. The American Agriculture Movement and the Call to Farm

1. Parity is a central idea and goal of many of the informants who participated in this study. Stock (1996:158) defines parity as "100 percent of the buying power of the agricultural sector during the good years between 1909 and 1914," an era during which it was believed that the value of agricultural products was essentially equal to industrial products (Dudley 2000:150). Mr. Gains mentions that "parity is equality." I believe he is referring to the notion that farmers should have the financial means to enable them to have the same access to goods and services that people in other professions enjoy—in other words to have access to the same standard of living.

2. The formulation of agricultural policy during the New Deal era in the U.S. is the focus of some debate. Some, such as Fitzgerald (2001) and Stock (1992), see the policy efforts of the U.S. Department of Agriculture continuing a statist tradition within agriculture: the promotion and creation of state-centered institutions that sought to advance technology and the continuing industrialization of American agriculture. These "high modernists" (Fitzgerald 2001) were relatively disconnected from the realities of rural life. Others, such as Gilbert (2003), provide an alternative perspective of New Deal agriculture, one that "stresses the agrarian intellectuals' work for family farmers, progressive reform and democratizing principles" (Gilbert 2003:214).

3. I am indebted to the work of Marty Strange for helping me make sense of the various issues associated with the changing structure of agriculture and its effects on farmers and local communities. I draw heavily from his work in this section.

4. It is interesting to note that the AAM's proposed solution to the crisis in family farming is a market solution, and directly opposed to the recommendation of the Center for Rural Affairs and some church advocacy groups to provide subsidies to farm families with insufficient income. This fact supports Dudley's (2003) assertion that, despite differing farm goals and practices and the existence of agrarian values, all farmers are well aware that they are producing for and have embraced capitalist markets. In my experience, what the AAM resented was not capitalism, but the injustice they perceived in its contemporary practice, especially government policies that favored large producers at the expense of family farmers.

5. The alliance's demand that the state be allowed to engage in any industry was amended to allow the state to engage in all except agriculture. This was a serious blow to the most radical agrarian faction of the group that met in Shawnee to draft a constitution for Oklahoma (Thompson 1986:79).

6. In a certain sense their concern over the Federal Reserve System is understandable. The contractionary monetary policy of the Federal Reserve under the Reagan Administration had a profound impact on interest rates; their unprecedented growth strongly affected the agricultural sector, which had become highly leveraged (Barnett 2003:166–167).

7. Stock (1996) argues that the activities of rural extremist organizations beginning in the 1960s cannot be viewed in isolation, but as part of a tradition of producer radicalism that predates the American Revolutionary War. The clues to making sense of the contradictory politics of these contemporary movements—which have included "contempt for the federal government, profound antiauthoritarianism, mockery of big business and finance, dedication to complete local control of the community and the desire to establish wilderness compounds" (Stock 1996:3)—lie in the particular configuration of five cultural processes that converge in rural areas. These include frontier life, class, race, gender, and evangelism. Stock (1996:7) argues that these five divisions of power in rural areas "create a special mix of contradictory experiences, impulses, ideologies, and actions ready to boil over into radical protest and collective violence in moments of economic, political, or cultural strain."

8. For this section, I drew heavily from an academic paper whose authorship I have not been able to establish. The document was given to me by an informant and remains in my files.

BIBLIOGRAPHY

Abu-Lughod, Lila. 1986. *Veiled Sentiments: Honor and Poetry in a Bedouin Society.*
Berkeley: University of California Press.

Abu-Lughod, Lila and Catherine A. Lutz. 1990. "Introduction: Emotion, Discourse, and the Politics of Everyday Life." In *Langage and the Politics of Emotion,* C. Lutz and L. Abu-Lughod, eds., 1–23. Cambridge: Cambridge University Press.

Adams, Jane. 1994. *The Transformation of Rural Life: Southern Illinois, 1890-1990.*
Chapel Hill: University of North Carolina Press.

American Agriculture Movement. n.d. *Who Understands Farmers and Farm Issues In Washington? Family Farmers.* Washington, D.C. In the author's files.

——. n.d. History of the American Agriculture Movement. In the author's files.

Arensberg, Conrad M. 1955. "American Communities." *American Anthropologist* 57:1143–1162.

Barlett, Peggy F. 1987. "The Crisis in Family Farming: Who Will Survive?" In *Farm Work and Fieldwork: American Agriculture in Anthropological Perspective*, M. Chibnik, ed., 29–57. Ithaca: Cornell University Press.

——. 1989. "Industrial Agriculture." In *Economic Anthropology*, S. Plattner, ed., 253–291. Stanford: Stanford University Press.

——. 1993. *American Dreams, Rural Realities: Family Farms in Crisis*. Chapel Hill: University of North Carolina Press.

Barlett, Peggy F. and Katherine Jewsbury Conger. n.d. "Three Visions of Masculine Success on American Farms." Manuscript in author's files.

Barnett, Barry J. 2003. "The U.S. Farm Financial Crisis of the 1980s." In *Fighting for the Farm: Rural America Transformed*, J. Adams, ed., 160–171. Philadelphia: University of Pennsylvania Press.

Behar, Ruth. 1995. "Introduction: Out of Exile." In *Women Writing Culture*, R. Behar and D. Gordon, eds., 1–29. Berkeley: University of California Press.

——. 1996. *The Vulnerable Observer: Anthropology that Breaks Your Heart*. Boston: Beacon Press.

Bellmon, Pat. 1993. *Spirit of '93: Stories of the Cherokee Strip Land Run*. Enid, OK: Cherokee Strip Centennial Foundation. Videocassette.

Belmonte, Laura. 2003. "Selling Capitalism: Modernization and U.S. Overseas Propaganda, 1945–1959." In *Modernization, Development, and the Globalization of the Cold War*, M. Latham, N. Gilman, M. Haefele, and D. Engerman, eds., 107–128. Amherst: University of Massachusetts Press.

Bennett, John W. 1982. *Of Time and the Enterprise: North American Family Farm Management in a Context of Resource Marginality*. Minneapolis: University of Minnesota Press.

Bourgois, Philippe. 1994. *In Search of Respect: Selling Crack in El Barrio*. Cambridge: Cambridge University Press.

Bowlin, John and Peter G. Stromberg. 1997. "Representations and Reality in the Study of Culture." *American Anthropologist* 9:123–134.

Brandes, Stanley. 1980. *Metaphors of Masculinity: Sex and Status in Andalusian Folklore*. Philadelphia: University of Pennsylvania Press.

Butler, Judith. 1996. *The Psychic Life of Power: Theories of Subjection*. Stanford: Stanford University Press.

Chodorow, Nancy. 1978. *The Reproduction of Mothering*. Berkeley: University of California Press.

——. 1999. *The Power of Feelings: Personal Meaning in Psychoanalysis, Gender and Culture*. New Haven: Yale University Press.

Cochrane, Willard W. 1993. *Development of American Agriculture: A Historical Analysis*. Minneapolis: University of Minnesota Press.

Collier, Jane. 1986. "From Mary to Modern Woman: The Material Basis of Marianismo and Its Transformation in a Spanish Village." *American Ethnologist* 13: 100–107.

——. 1997. *From Duty to Desire: Remaking Families in a Spanish Village*. Princeton: Princeton University Press.

Conway, Jill Ker. 1993. *True North: A Memoir*. New York: Alfred A. Knopf.

Debo, Angie. 1940. *And Still the Waters Run: The Betrayal of the Five Civilized Tribes*. Princeton: Princeton University Press.

Delaney, Carol. 1991. *The Seed and the Soil: Gender and Cosmology in Turkish Village Society*. Berkeley: University of California Press.

——. 1998. *Abraham on Trial: The Social Legacy of a Biblical Myth*. Princeton: Princeton University Press.

Derby, Jill. 1989. "Women's Roles on Economically Depressed Nevada Ranches." In *Food and Farm: Current Debates and Policies*, C. H. Gladwin and K. Truman, eds., 321–345. Lanham: University Press of America.

Dirks, Nicholas B, Geoff Eley, and Sherry B. Ortner. 1994. "Introduction." In *Culture/Power/History: A Reader in Contemporary Social Theory*, N. B. Dirks, G. Eley, and S. B. Ortner, eds., 3–45. Princeton: Princeton University Press.

Dudley, Kathryn Marie. 2000. *Debt and Dispossession: Farm Loss in America's Heartland*. Chicago: University of Chicago Press.

——. 2003. "The Entrepreneurial Self: Identity and Morality in a Midwestern Farming Community." In *Fighting for the Farm: Rural America Transformed*, J. Adams, ed., 175–191. Philadelphia: University of Pennsylvania Press.

Dyer, Joel. 1996. "Remember the Farm Crisis?" *Utne Reader*, November-December.

Eley, Geoff. 1994. "Nations, Publics and Political Cultures: Placing Habermas in the Nineteenth Century." In *Culture/Power/History: A Reader in Contemporary Social Theory*, N. B. Dirks, G. Eley and S. B. Ortner, eds., 297–335. Princeton: Princeton University Press.

Errington, Frederick. 1990. "Rock Creek Rodeo: Excess and Constraint in Men's Lives." *American Ethnologist* 17:623–645.

Escobar, Arturo. 1995. *Encountering Development: The Making and Unmaking of the Third World*. Princeton: Princeton University Press.

Faludi, Susan. 1999. *Stiffed: The Betrayal of the American Man*. New York: William Morrow and Company.

Farmer, Paul. 1999. *Infections and Inequalities: The Modern Plagues*. Berkeley: University of California Press.

——. 2003. *Pathologies of Power: Health, Human Rights and the New War on the Poor*. Berkeley: University of California Press.

Fink, Deborah. 1992. *Agrarian Women: Wives and Mothers in Rural Nebraska, 1880–1940*. Chapel Hill: University of North Carolina Press.

Fitchen, Janet M. 1991. *Endangered Spaces, Enduring Places: Change, Identity and Survival in Rural America*. Boulder: Westview Press.

Fitzgerald, Deborah. 2001. "Accounting for Change: Farmers and the Modernizing State." In *The Countryside in the Age of the Modern State: Political Histories of Rural America*, C. M. Stock and R. D. Johnson, eds., 189–212. Ithaca: Cornell University Press.

——. 2003. *Every Farm a Factory: The Industrial Ideal in American Agriculture*. New Haven: Yale University Press.

Foucault, Michel. 1978. *The History of Sexuality, Volume I: An Introduction*. New York: Vintage Books.

Fraser, Nancy. 1985. "What's Critical about Critical Theory? The Case of Habermas and Gender." *New German Critique* 35:97–131.

Fricker, Miranda. 1991. "Reason and Emotion." *Radical Philosophy* 57:14–19.

Friedberger, Mark. 1988. *Farm Families and Change in Twentieth-Century America*. Lexington: University Press of Kentucky.

Gibson, Arrell Morgan. 1981. *Oklahoma: A History of Five Centuries*. Norman: University of Oklahoma Press.

Gilbert, Jess. 2003. "Low Modernism and the Agrarian New Deal: A Different Kind of State." In *Fighting for the Farm: Rural America Transformed*, J. Adams, ed., 129–146. Philadelphia: University of Pennsylvania Press.

Gilmore, David D. 1990. *Manhood in the Making: Cultural Concepts of Masculinity*. New Haven: Yale University Press.

Ginsburg, Faye D. 1989. *Contested Lives: The Abortion Debate in an American Community*. Berkeley: University of California Press.

Gladwin, Christina H. 1989. "On the Division of Labor Between Economics and Economic Anthropology." In *Economic Anthropology*, S. Plattner, ed., 397–425. Stanford: Stanford University Press.

Goldschmidt, Walter. 1978. *As You Sow: Three Studies in the Social Consequences of Agribusiness*. Montclair, NJ: Allanheld, Osmun, and Company.

Gregor, Thomas. 1985. *Anxious Pleasures: The Sexual Lives of an Amazonian People*. Chicago: University of Chicago Press.

Gupta, Akhil. 1998. *Postcolonial Developments: Agriculture in the Making of Modern India*. Durham: Duke University Press.

Harding, Susan Friend. 1984. *Remaking Ibieca: Rural Life in Aragon Under Franco*. Chapel Hill: University of North Carolina Press.

Heffernan, Judith Bortner. 1989. "Keynote Address." Presented at the Globalization of the Food System Conference, Freedom, Oklahoma, March.

Hill, Luther B. 1909. *A History of the State of Oklahoma*. Chicago: Lewis Publishing Company.

Human, Jeffrey and Cathy Wasem. 1991. "Rural Mental Health in America." *American Psychologist* 46:232–239.

Jurich, Anthony P. and Candyce S. Russell. 1987. "Family Therapy with Rural Families in a Time of Crisis." *Family Relations* 36:364–367.

Kimmel, Michael. 1996. *Manhood in America: A Cultural History*. New York: The Free Press.

Kondo, Dorinne. 1986. "Dissolution and Reconstitution of Self: Implications for Anthropological Epistomology." *Cultural Anthropology* 1:74–96.

Kundera, Milan. 1981. *The Book of Laughter and Forgetting*. New York: Penguin Books.

Lamar, Howard R. 1993. "The Creation of Oklahoma: New Meanings for the Oklahoma Land Runs." In *The Culture of Oklahoma*, H. F. Stein and R. F. Hill, eds., pp. 29–47. Norman: University of Oklahoma Press.

Lee, Mark. 1989. "Study Shows Suicide High Among Farmers." *Tulsa World*, 12 September 1989, A1.

Levitas, Daniel and Leonard Zeskind. 1987. "The Farm Crisis and the Radical Right." In *Renew the Spirit of My People: A Handbook for Ministry in Times of Rural Crisis*, K. Schmidt, ed., 23–30. Des Moines: Prairiefire Rural Action, Inc.

Lutz, Catherine. 1988. *Unnatural Emotions: Everyday Sentiments on a Micronesian Atoll and their Challenge to Western Theory*. Chicago: University of Chicago Press.

Mead, Margaret. 1942. *And Keep Your Powder Dry: An Anthropologist Looks at America*. New York: William Morrow and Company.

Middleton, Peter. 1992. *The Inward Gaze: Masculinity and Subjectivity in Modern Culture*. New York: Routledge.

Mitchell, Timothy. 2000. "Introduction." In *Questions of Modernity*, T. Mitchell, ed., xi–xxvii. Minneapolis: University of Minnesota Press.

Morgan, David R., Robert E. England, and George G. Humphreys. 1991. *Oklahoma Politics and Policies: Governing the Sooner State*. Lincoln: University of Nebraska Press.

Morgan, H. Wayne and Anne Hodges. 1977. *Oklahoma: A History*. Norman: University of Oklahoma Press.

Morgen, Sandra. 1992. "'It was the best of times, it was the worst of times': Work Culture in Feminist Health Clinics." Article in the author's files.

Murdock, Steve H. and F. Larry Leistritz. 1988. *The Farm Financial Crisis: Socioeconomic Dimensions and Implications for Producers and Rural Areas*. Boulder: Westview Press.

Murray, J. Dennis and Peter A. Keller. 1991. "Psychology and Rural America: Current Status and Future Directions." *American Psychologist* 46:220–231.

Nikolitch, Radoje. 1969. "Family-Operated Farms: Their Compatibility with Technological Advance." *American Journal of Agricultural Economics* 51:530–545.

O'Brien, Edna. 2000. *Wild Decembers*. Boston: Houghton Mifflin and Company.

Oklahoma State Department of Health. 1991. *Farm Related Deaths*. Oklahoma City: Injury Epidemiology Division.

Oklahoma State University Cooperative Extension Service. 1991. *Oklahoma Helping Hand: A Program to Assist Financially Stressed and Dislocated Farmers and Rural Families*. Stillwater: Department of Agriculture, Oklahoma State University.

Ortner, Sherry B. 1974. "Is Female to Male as Nature is to Culture?" In *Woman, Culture and Society*, M. Rosaldo and L. Lamphere, eds., 67–87. Stanford: Stanford University Press.

Rainey, George. 1933. *The Cherokee Strip*. Guthrie, OK: Co-Operative Publishing Company.

Reddy, William M. 1997. *The Invisible Code: Honor and Sentiment in Postrevolutionary France, 1814–1848*. Berkeley: University of California Press.

Ritchie, Mark. 1979. *The Loss of Our Family Farms: Inevitable Results or Conscious Policy?* Minneapolis and Hampton, Iowa: Center for Rural Studies and the U.S. Farmers Association.

Rogers, Susan Carol. 1991. *Shaping Modern Times in Rural France: The Transformation and Reproduction of an Aveyronnais Community*. Princeton: Princeton University Press.

Rojas-Smith, Lucia, Eric E. Ramirez, and Richard Perry. 1991. *Helping Hands Rural Mental Health Project Report*. Enid, OK: Rural Health Projects, Inc.

Rosaldo, Michelle Z. 1974. "Woman, Culture and Society: A Theoretical Overview." In *Woman, Culture and Society*, M. Rosaldo and L. Lamphere, eds., 17–42. Stanford: Stanford University Press.

———. 1984. "Toward an Anthropology of Self and Feeling." In *Culture Theory: Essays on Mind, Self, and Emotion*, R. A. Shweder and R. A. LeVine, eds., 137–157. Cambridge: Cambridge University Press.

Rosaldo, Renato. 1989. *Culture and Truth: The Remaking of Social Analysis*. Boston: Beacon Press.

Rosenblatt, Paul C. 1990. *Farming is in Our Blood: Farm Families in Economic Crisis*. Ames: Iowa State University Press.

Rotundo, E. Anthony. 1993. *American Manhood: Transformations in Masculinity from the Revolution to the Modern Era*. New York: Basic Books.

Said, Edward. 1999. *Out of Place: A Memoir*. New York: Alfred A. Knopf.

Salamon, Sonya. 1992. *Prairie Patrimony: Family, Farming and Community in the Midwest*. Chapel Hill: University of North Carolina Press.

———. 2003. *Newcomers to Old Towns: Suburbanization of the Heartland*. Chicago: University of Chicago Press.

Scheper-Hughes, Nancy. 1992. *Death Without Weeping: The Violence of Everyday Life in Brazil*. Berkeley: University of California Press.

Schertz, Lyle P., et al. 1979. *Another Revolution in U.S. Farming?* U.S. Department of Agriculture, Economic Research Service, Agricultural Economic Report no. 411. Washington, D.C.: U.S. Government Printing Office.

Schneider, David M. 1968. *American Kinship: A Cultural Account*. Chicago: University of Chicago Press.

Sloan, Tod. 1996. *Damaged Life: The Crisis of the Modern Psyche*. New York: Routledge.

Smiley, Jane. 1991. *A Thousand Acres*. New York: Fawcett Columbine.

Smith-Rosenberg, Carroll. 1985. *Disorderly Conduct: Visions of Gender in Victorian America*. New York: Oxford University Press.

Spindler, George and Louise Spindler. 1983. "Anthropologists View American Culture." *Annual Review of Anthropology* 12:49–78.

Stein, Howard F. 1987a. "Farmer and Cowboy: The Duality of the Midwestern Male Ethos—a Study in Ethnicity, Regionalism, and National Identity." In *From Metaphor to Meaning: Papers in Psychoanalytic Anthropology*, H. F. Stein and M. Apprey, eds., 178–227. Charlottesville: University of Virginia Press.

———. 1987b. "The Annual Cycle and the Cultural Nexus of Health Care Behavior Among Oklahoma Wheat Farmers." In *From Metaphor to Meaning: Papers in Psychoanalytic Anthropology*, in H. F. Stein and M. Apprey, eds., 156–177. Charlottesville: University of Virginia Press.

———. 1992. "Cultural Issues In Oklahoma Rural Medicine: A Place to Begin (Part II)." *AHEC News* 3:7–8.

———. 1996. *Prairie Voices: Process Anthropology in Family Medicine*. Westport: Bergin and Garvey.

Stein, Howard F. and Gary L. Thompson. 1992. "The Sense of Oklahomaness: Contributions of Psychogeography to the Study of American Culture." *Journal of Cultural Geography* 11:63–91.

Steinbeck, John. 1989. *The Grapes of Wrath: The Fiftieth Anniversary Edition*. New York: Viking.

Stewart, Kathleen. 1996. *A Space on the Side of the Road: Cultural Poetics in an "Other" America*. Princeton: Princeton University Press.

Stock, Catherine McNicol. 1992. *Main Street in Crisis: The Great Depression and the Old Middle Class on the Northern Plains*. Chapel Hill: University of North Carolina Press.

———. 1996. *Rural Radicals: Righteous Rage in the American Grain*. Ithaca: Cornell University Press.

Strange, Marty. 1988. *Family Farming: A New Economic Vision*. Lincoln and San Francisco: University of Nebraska Press and the Institute for Food and Development Policy.

Stromberg, Peter G. 1986. *Symbols of Community: The Cultural System of a Swedish Church*. Tucson: University of Arizona Press.

Summers, Mary. 2001. "From the Heartland to Seattle: The Family Farm Movement of the 1980s and the Legacy of Agrarian State Building." In *The Countryside in the Age of the Modern State: Political Histories of Rural America*, C. M. Stock and R. D. Johnston, eds., 304–325. Ithaca: Cornell University Press.

Sweeny, John. 1999. "Making Globalization Work for All." *Call to Action Spirituality/Justice Reprint*, June.

Terdiman, Richard. 1985. *Discourse/Counter-Discourse: The Theory and Practice of Symbolic Resistance in Nineteenth-Century France*. Ithaca: Cornell University Press.

Thompson, John. 1986. *Closing the Frontier: Radical Response in Oklahoma, 1889–1923*. Norman: University of Oklahoma Press.

Turner, Alvin O. and Vicky L. Gailey. 1998. "Enid's Golden Era, 1916–1941." *The Chronicles of Oklahoma* 76:116–139.

U.S. Department of Agriculture. 1980. *A Time to Choose: Summary Report on the Structure of Agriculture*. Washington, D.C.: U.S. Government Printing Office.

Visweswaran, Kamala. 1994. *Fictions of a Feminist Ethnography*. Minneapolis: University of Minnesota Press.

Wagenfeld, Morton O. 1990. "Mental Health and Rural America: A Decade Review." *Journal of Rural Health* 6:507–522.

Weedon, Chris. 1988. *Feminist Practice and Poststructuralist Theory*. Oxford: Basil Blackwell.

Wells, Miriam J. 1996. *Strawberry Fields: Politics, Class and Work in California Agriculture*. Ithaca: Cornell University Press.

Weston, Kath. 1998. *Long Slow Burn: Sexuality and Social Science*. New York: Routledge.

Williams, Raymond. 1977. *Marxism and Literature*. Oxford: Oxford University Press.

——. 1979. *Politics and Letters: Interviews with New Left Review*. London: New Left Press.

Yanagisako, Sylvia Junko. 2002. *Producing Culture and Capital: Family Firms in Italy*. Princeton: Princeton University Press.

Yanagisako, Sylvia Junko and Jane Fishburne Collier. 1987. "Toward a Unified Analysis of Gender and Kinship." In *Gender and Kinship: Essays Toward a Unified Analysis*, J. F. Collier and S. J. Yanagisako, eds., 14–50. Stanford: Stanford University Press.

Yanagisako, Sylvia and Carol Delaney. 1995. "Naturalizing Power." In *Naturalizing Power: Essays in Feminist Cultural Analysis*, S. Yanagisako and C. Delaney, eds., 1–22. New York: Routledge.

INDEX

Abraham, story of, 159–160
Abu-Lughod, L., 62, 63
accidents, farm, 2, 53
acquisitiveness, American, 110
Adams, J., 96
AG-LINK hotline, 2, 3, 52, 143, 158
agrarianism, women in, 200–201n. 3
agricultural exports, in global market, 96
agricultural industry, in 1980s, 8
agricultural policy: centralized development of, 170; and family farmers, 138–139

agriculture: boom in, 93–94; "progressive," 176; and rural social change, 23; structure of, 130, 134, 138; transformation of, 10. *See also* American Agriculture Movement; farming, family-centered
agriculture, industrial, 87; and changing cultural discourses, 120–121; compared with family-farm agriculture, 130; farm management in, 102; hegemony of, 175; management science in, 107; marginalization associated with, 119–120; and